W9-BZO-093

A Field Guide to the **Tiger Beetles**
of the United States and Canada

A Field Guide to the

BY DAVID L. PEARSON

C. BARRY KNISLEY

DANIEL P. DURAN

CHARLES J. KAZILEK

Tiger Beetles

of the United States and Canada

Identification,

Natural History,

and Distribution

of the

Cicindelinae

2nd Edition

OXFORD
UNIVERSITY PRESS

OXFORD
UNIVERSITY PRESS

Oxford University Press is a department of the University of
Oxford. It furthers the University's objective of excellence in research,
scholarship, and education by publishing worldwide.

Oxford   New York
Auckland   Cape Town   Dar es Salaam   Hong Kong   Karachi
Kuala Lumpur   Madrid   Melbourne   Mexico City   Nairobi
New Delhi   Shanghai   Taipei   Toronto

With offices in
Argentina   Austria   Brazil   Chile   Czech Republic   France   Greece
Guatemala   Hungary   Italy   Japan   Poland   Portugal   Singapore
South Korea   Switzerland   Thailand   Turkey   Ukraine   Vietnam

Oxford is a registered trademark of Oxford University Press
in the UK and certain other countries.

Published in the United States of America by
Oxford University Press
198 Madison Avenue, New York, NY 10016

© Oxford University Press 2015

Cataloging-in-Publication data is on file at the Library of Congress
ISBN 978–0–19–936716–0 (hbk.); 978–0–19–936717–7 (pbk.)

Contents

Preface

Within months after the appearance of the first edition of this field guide, we began to receive suggestions for changes, corrections of errors, and new observations of seasonality, range extensions, and biology. The recent publication of several major taxonomic revisions of tiger beetles also changed many of the concepts we had used in the first edition. The critical mass of these changes finally reached a point where we thought these updates needed to be included in a second revision.

The journal *CICINDELA* continues to be a great help in providing new information about tiger beetles, but more and more journals in America, Europe, Australia, Asia, and South America have become important for both professionals and a growing number of amateurs to publish their observations. We hope that the increasing number of regional field guides to tiger beetles across the world will reflect our experience with the first edition of *The Field Guide to the Tiger Beetles of the United States and Canada*. Many young people, established naturalists, and outdoor enthusiasts, when given the means to recognize and easily identify tiger beetles, become tiger beetle aficionados. These enthusiastic amateurs advance our knowledge of nature around the world to a level that would be impossible for the professionals to do on their own.

We strove to write the first edition and now this second edition of the guide in a pleasant and comprehensible style that limits scientific jargon and concepts. Even more basic, however, was resolving simple problems such as developing an acceptable list of common names. Over almost a year, an ad hoc committee of ten stalwart volunteers sifted through suggestions and debated details. We even enlisted the entire readership of the journal *CICINDELA* to vote their favorite common names and provide suggestions for alternatives. The resulting selection was a cautious consensus, and these are the common names we use here.

We also knew that the color identification plates would be essential to this field guide meeting its goals. These plates needed to be artistically pleasing as well as scientifically accurate if the field guide was to be useful and attract many new enthusiasts. Charles Kazilek worked diligently and with great passion on the twenty-five species identification color plates that grace this field guide. Without his work and partnership in this effort, the field guide would not have been possible. With real specimens and the magic of digital photography and Adobe PhotoShop, he painted the illustrations with pixels in all their glory and color but without sacrificing identification details.

To write this field guide, we relied not only on our combined eighty-five years of experience studying tiger beetles, but also unabashedly drafted the help of many of our colleagues. They generously shared their knowledge and interest in tiger beetles, so that we could assemble as complete a picture as possible of the identification, distribution, natural history, and habitat details of the 116 species of tiger beetles occurring on the continent. Those who provided indispensable information and advice include John Acorn, Giff Beaton, Chris R. Brown, Matt Brust, David Brzoska, Gary Budyk, Deannna Dodgson, Richard Freitag, Henri Goulet, Alan Harvey, David Herrmann, Wyatt Hoback, Ronald Huber, Michael Kippenhan, Ron Lyons, Ted MacRae, Jonathan Mawdsley, Bob Pape, Will Richardson, David Roemer, Steve Roman, W. Dan Sumlin II, Steve Spomer, and Gretchen Waggy. Photographs of live tiger beetles in the field were generously supplied by Giff Beaton, Christopher Brown, Matt Brust, Kevin Fielding, Eva Furner, Ted MacRae, Fred Pfeifer, David Rogers, David Roemer, and Stephen Spomer. Alan Harvey and Michael Kippenhan produced several of the line drawings. Cameras and technical assistance for photo imaging were supplied by Nico Franz and the ASU Hasbrouck Insect Collection.

Finally, but certainly not least, we thank our wives, Nancy Pearson, Peggy Knisley, Melissa Duran, and Sally Kazilek, for their patience and support over these many years.

David L. Pearson
School of Life Sciences
Arizona State University
Tempe, Arizona 85287-4501 USA

C. Barry Knisley
Department of Biology
Randolph-Macon College
Ashland, Virginia 23005-5505 USA

Daniel P. Duran
Department of Biology
Drexel University
Philadelphia, Pennsylvania 19104-2875 USA

Charles J. Kazilek
Vice Provost
Office of University Provost
Arizona State University
Tempe, Arizona 85287-7805 USA

A Field Guide to the **Tiger Beetles**
of the United States and Canada

The Magic of Tiger Beetles

Hundreds, possibly thousands, of otherwise normal people are passionate about an intriguing group of insects called tiger beetles (cicindelids). These cicindelophiles or tiger beetlists can be found swinging insect nets to collect wary tiger beetles from sandy pine forests in Florida to alpine meadows in the Canadian Rocky Mountains. Others might spend hours crawling on elbows and knees in the piercing heat of an Arizona grassland for close-up photos of a ruby-red species. A few sit patiently on the white sands of a New Jersey ocean beach taking notes about the mating behavior of a long-legged tiger beetle, whose population is listed by the U.S. government as an endangered species.

These activities and others unite a diverse group of tiger beetle enthusiasts, only a few of whom are professional entomologists. Tiger beetles elicit something more than a routine response to the necessities of employment. Note the highly enthusiastic and curious behavior of some lawyers, dentists, prison guards, railroad workers, computer engineers, and a bevy of amateurs who have found a life-long hobby in this beetle family. If there is some kind of magic emitted by tiger beetles, what is it, and how has it captured the interest of so many people regardless of whether they are being paid for it?

It is a question we are asked frequently, and it is not one that can be easily or logically explained. All we know is that you have to be exposed to a tiger beetle to discover if this passion lies inside you, too. For some, it is seeing a group of spectacularly colored specimens pinned in perfect rows in a glass-topped insect drawer. For many others, it is noticing a live tiger beetle for the first time on a forested path, along a river bank, or in some other place that may have been taken for granted until this moment. We know that this interest in tiger beetles is not mystical, but if you talk to tiger beetle aficionados about their hobby or study, many of them will not be able to explain the source of what the uninitiated may see as a mania.

A large part of the reason we decided to write this field guide was to persuade others that tiger beetles are special. The book should facilitate an interest in tiger beetles by amateurs and professionals alike because it serves as the first combined source of illustrated identification, natural history, and distribution for all the species and subspecies of tiger beetles known to occur in the United States and Canada.

Color plates of each species, subspecies, and many intermediate forms will allow readers to use this book as a guide that permits a direct comparison of beetles in the field or a mounted specimen in a collection with the pictures. For others who might consider this method of comparing specimens with pictures too "hit-and-miss," we provide simple keys. The couplets in these keys are generously supplied with line drawings of pertinent morphological features to reinforce the substance of distinguishing descriptions and terms. Although the keys concentrate primarily on adult characters, an abbreviated key to the larval forms also is provided to the level of genera. In addition, detailed species accounts are provided, and they include information on both adult and larval behavior, habitat, similar species, and other data useful for recognizing species and subspecies and when to expect to find them in the field.

More than 2700 species of tiger beetles have been described to date, and they are found all over the world's land surface except Antarctica, the Arctic north of 65° latitude, and some isolated oceanic islands like Hawaii and the Maldives. In elevation, they range from about 3500 m above sea level to 220 m below sea level. In North America north of Mexico, there are 116 species, which have been divided into an additional 153 recognized subspecies or geographically distinct races. Detailed studies of their natural history, population dynamics, communities, patterns of worldwide species richness, and taxonomy of particular subgroups have produced much information. Tiger beetles, as a result, are among the most widely investigated groups of insects, especially in terms of their ecology and geographic distribution. Details of the existing wealth of biological information have been synthesized in *Tiger Beetles: The Evolution, Ecology, and Diversity of the Cicindelids,* by David L. Pearson and Alfried Vogler, which serves as a companion book to this field guide. Four regional publications also emphasize identification and natural history of tiger beetles in parts of North America: *The Biology of Tiger Beetles and a Guide to the Species of the South Atlantic States,* by C. Barry Knisley and Thomas Schultz; *Northeastern Tiger Beetles: A Field Guide to Tiger Beetles of New England and Eastern Canada,* by Jonathan Leonard and Ross Bell; *Tiger Beetles of Alberta: Killers on the Clay, Stalkers on the Sand,* by John Acorn; and *Tiger Beetles: A Field Guide and Identification Manual for Florida and Eastern U.S.,* by Paul Choate. For distribution of tiger beetle species, we used maps in these publications as well as many articles in the quarterly journal *CICINDELA.* Numerous websites are available electronically for recent tiger beetle information at a local or state scale, and we perused them frequently for pertinent updates to make this field guide as current as possible.

In this day of genomic studies, stem cells, and molecular clocks, a field guide to a family of beetles may seem antediluvian to some. However, it is a small jump from cellular studies to placing names and associating distribution and natural history with beetles such as these. Tiger beetles have already proved themselves to be ideal models for understanding many other parts of the biosphere that are themselves so complicated that they hinder advances in our knowledge. Studies of tiger beetle mitochondrial and nuclear DNA have begun to open doors of comprehension that extend to many other organisms. New insights into tiger beetle taxonomy and evolutionary relationships have been based largely on recent studies of molecules. We include them in this second edition.

Investigations of reflectance mechanisms of tiger beetle colors, ultrasound hearing, visual systems, exceptional genetic anomalies, and patterns of species distributions across continents have all been pioneered using tiger beetles. However, without names to attach to species and without basic knowledge of natural history and distribution, these and other sophisticated studies would be much more difficult, if not impossible. For those who need less justification for a field guide, basic procedures, such as putting names on living or dead specimens, can be a means of deepening a relationship between observer and beetle, as well as facilitating communication among enthusiasts.

The first edition of this field guide enabled so many new enthusiasts to contribute new data and observations that within six years it became obsolete. Because of this cadre of new tiger beetle devotees, our knowledge of distribution for many tiger beetles has changed dramatically; behavior and ecology of adults and larvae have been newly described or supplemented; and at least eight new species and numerous subspecies have been described from North America since the first edition was published. We now need to bring the knowledge of tiger beetles up to date, but our sincerest hope is that even more new tiger beetle fanatics will be infected with this second edition of the field guide and soon make it equally as obsolete.

To slow this inevitable tide of obsolescence, we will be partnering with the Ask A Biologist program at Arizona State University. The website is visited by millions each year, and now you can point your browser to the new tiger beetle section, http://askabiologist.asu.edu/tiger-beetle-watcher, to keep yourself up to date or share the link with a new enthusiast. This section has been developed to complement the field guide by providing recent observations of behavior and natural history, current reports of range extensions (or reductions), conservation efforts, taxonomic changes and explanations, and references to the latest popular and scientific articles on all aspects of tiger beetles. There are also sections for kids and newbies to engage a new generation of tiger beetle enthusiasts.

How to Recognize a
Tiger Beetle

Tiger beetles form a distinct group of species within the order Coleoptera, which contains all the beetles. They are now included by most authors as a distinct subgroup (subfamily Cicindelinae, supertribe Cicindelitae, or tribe Cicindelini) within the family of ground beetles called Carabidae. Before the availability of modern molecular techniques and phylogenetic analyses, they were considered by some entomologists to form their own family, Cicindelidae. They share many characters in common with and are most closely related to the predaceous ground beetles (Carabidae), predaceous diving beetles (Dytiscidae), whirligig beetles (Gyrinidae), and crawling water beetles (Haliplidae). These five families and a few others are placed together as a suborder called Adephaga. Fourteen genera of tiger beetles occur in North America, and several characters in combination reliably distinguish tiger beetles from all other groups of Adephaga: (1) long, sickle-shaped mandibles; (2) simple teeth arranged along the inner side of the mandible with a compound (molar-like) tooth on the inner base of each mandible; (3) long, thin antennae with 11 segments and attached to the head between the eye and base of the mandible; (4) long body form with eyes and head together wider than the thorax; (5) long thin running legs; (6) tunnel-building behavior of the larvae; and (7) peculiar forward-facing sets of hooks on the backs of the larvae.

Most adult tiger beetles species look remarkably similar in body shape, proportions, and behavior. They vary primarily in size, color, and shape of markings. In North America, the smallest species is barely 7 mm long, while the largest is greater than 70 mm. Many species are dull black, but some species are brilliantly emblazoned with bright green, violet, blue, red, and yellow. Most species have streamlined bodies and long slender legs for fast sprinting across the ground or vegetation. The prominent eyes of these visual hunters are usually so large that they make the head wider than the relatively narrow thorax. Adults have transparent hind wings that are usually folded and hidden under the hard front wings, the elytra. In flight, these elytra conveniently open forward to allow the flight wings underneath to unfold and extend out to the sides. The wings are used for short and low escape flights from predators. A few species, however, use their wings for long-range dispersal, while a few others have lost these flight wings and are Earthbound.

On the ground, where most species spend their lives, tiger beetle adults typically run in short but fast spurts interspersed with brief stops. The stops are necessary because they run so fast (up to 8 kph) that after a while their brains become overwhelmed by sensory input and stop processing the changing light images. Their world becomes a blur, and they cannot see their prey. During these stops, they restore a clear image.

If it sees a potential prey item, such as an ant, a small spider, or a fly, the tiger beetle quickly turns in that direction and waits for another movement. The tiger beetle then runs the prey down and, if successful, grabs it with its long, thin, sickle-shaped mandibles. These mandibles are used to chew the prey into a puree. The beetle's mandibular glands near the base of each mandible release enzymes that begin the digestion process. The fluid is conducted from the gland to the mandibular tip and teeth via a groove. This chewing tobacco-like substance is also used in defense, and anyone who has collected tiger beetles with a light-colored net bag soon has a mesh with brown spots. A collector who is unlucky enough to have a larger tiger beetle mandible break the skin of his or her thumb also knows the sting of this fluid.

The larvae are peculiar among beetles. They all are designed for life in a narrow burrow. As a result, even though the adults may be nocturnal or diurnal, long and thin, or short and wide, the larvae are white and grub-like (Fig. 2.1), with much of the outer covering of their bodies membranous.

A dark armored capsule covers the head, and scattered dark plates are especially noticeable on the top of the thorax (pronotum). They have a large head with up to six small eyes on top and formidable mandibles

Figure 2.1 Side view of tiger beetle larva at the mouth of its burrow.

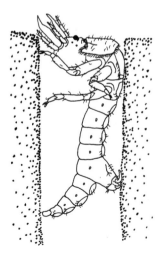

Figure 2.2 Head of larval Eastern Beach Tiger Beetle (*Habroscelimorpha dorsalis*) at mouth of its burrow. Photograph by G. Beaton.

underneath. A particularly striking feature is on the larva's lower back, which includes a prominent hump with two pairs of large hooks that face forward. The larvae, like adult tiger beetles, are carnivorous, but unlike the adults the larvae wait for prey to come to them. Each larva positions itself at the top of a long burrow with its head and thorax flush to the substrate surface and exactly filling the diameter of the burrow entrance. Larval burrows, depending on the species, can be on flat soil, vertical clay banks, forest leaf litter, or, for a few tropical taxa, even in rotted wood of branches and twigs. When a prey item approaches the burrow entrance closely, the larva extends its body, anchored by the back hooks in the side of the tunnel, and quickly reaches out backward to grab the prey in its powerful mandibles. The larva then pulls the struggling prey down into the depths of its burrow and dispatches it with a few mighty bites.

Because the head and thorax are usually the same color and texture as the surrounding soil surface, most larvae are hard to see as they wait at the top of the burrows (Fig. 2.2). Their reaction to danger is to retreat immediately away from the mouth of their burrows; thus, their presence is made obvious only when a black hole suddenly appears where before there was none.

Body Parts

Our keys for identifying tiger beetle species use simple language and are illustrated wherever possible. However, some basic knowledge of anatomy is necessary, and we include here a simple overview of adult and larval structural features. These illustrations and brief explanations serve best as a reference or kind of dictionary for translating otherwise arcane terms.

Adults

The hard armor-like skin (cuticle) that covers the adult tiger beetle is critical for survival, but it is also very useful in identification. The outermost layer (epicuticle) has patterns of tiny pits, larger punctures, ridges, and undulations called microsculpture. The differences in these patterns of microsculpture are frequently used to distinguish tiger beetle species and genera. The cuticle is laminated with layers of melanin pigment and translucent waxes that alternately reflect and pass light. The distance between these alternating layers produces a broad range of metallic colors through reflectance and interference. The degree of uniformity of the cuticular reflector determines the purity of reflected color. Highly sculptured and nonuniform reflectors produce a broad blend of colors of different wavelengths reflected at various angles from different locations. This type of integument gives rise to dull green or brown similar to those made by pigments in other insects. In bright iridescent species, the cuticular sublayers are more uniform, and the surface is relatively smooth. In all-black species, like those of the genera of Night-stalking Tiger Beetles (*Omus*) and Giant Tiger Beetles (*Amblycheila*), the melanin is deposited in relatively thick and disorganized patterns that absorb most light. In other species, parts of the integument have no pigment deposited, and these areas are pale yellow or white.

The most frequently studied anatomical features of adult tiger beetles include the head (Fig. 2.3), where the distinctive characters include long thread-like (filiform) antennae with 11 segments (color, distribution of hair-like setae, relative overall length) used primarily as tactile sense organs; mandibles (relative length, number, and position of "teeth") used for capturing and processing prey and in males for grasping females during mating; upper lip or labrum (color, length–width ratio, number and position of "teeth," number and position of setae) used with the mandibles to help grasp and process prey; labium (presence and position of setae, relative length, and color of the segments of its finger-like palpi); and maxillae (relative length and color of the segments of its finger-like palpi) used to manipulate and analyze the quality of food items. Other important parts of the head include the compound eyes (degree of bulging, relative size), the complexity of surface microsculpturing and depth of grooves (rugae) between the eyes and other parts of the head, as well as the distribution of white, hair-like, or thick and flattened setae that may function as sense organs and/or insulation against hot surfaces.

Various adult eye sizes and shapes produce variable areas of stereoscopic (three-dimensional) vision. Nocturnal adults such as those of Giant

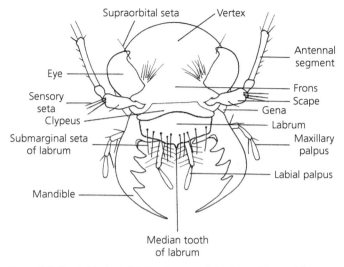

Figure 2.3 Head of typical adult tiger beetle and structures often used for identification.

Tiger Beetles, Night-stalking Tiger Beetles, and Metallic Tiger Beetles have small, relatively flat eyes compared with the bulbous eyes of genera active during the light of day.

On the thorax (Fig. 2.4), the most frequent distinctions are found in the proportions of the thorax (rectangular, square, or elongate), its shape as viewed from above (cylindrical, parallel, rounded, trapezoid, and so on), the texture (shiny metallic or dull) and color of the upper surface (pronotum), and the patterns or absence of setae on its side and lower or upper surfaces.

In back of the pronotum (Fig. 2.4), the hardened elytra (modified front wings) that cover the flight (hind) wings and top of the abdomen are probably the most commonly used identification characters. These elytra are spread and rotated forward in flight, where they may function as airfoils but do not flap. The elytral surface texture or microsculpturing can include large individual punctures (foveae), patterns of small pits (punctation), grooves (rugae), smooth (impunctate) areas, or undulations and tiny saw-like teeth (microserrations) and spines on the rear edge of the elytra. Also important are the shape of the elytra from above (parallel-sided, rounded, oval, and the like), their profile as viewed from the side (domed, flattened), their dark background color and texture (shiny metallic, dull), and the pattern, color and position of spots, lines, and stripes (maculations) or their absence (immaculate).

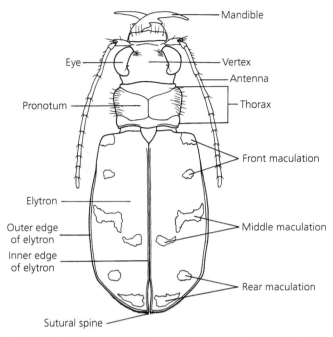

Figure 2.4 Top view of typical adult tiger beetle body and structures often used for identification.

The flight wings are membranous with a distinctive framework of thickened ridges called veins. Modifications of tiny structures allow for a triple folding so that the flight wings can be stored completely under the elytra. In some flightless species, these flight wings are shrunk or even totally absent and with permanently fused elytra.

The prominent legs (Fig. 2.5) of tiger beetles are thin and long for fast running. Leg color, positions of setae, and relative size (both to the overall size of the tiger beetle as well as to other segments on the same leg) are sometimes important as identifying characters. Males of all species have white pads of long curved setae on the feet (tarsi) of the front legs.

The form and shape of the male mating structures in tiger beetles are well known and have been used extensively in taxonomic comparisons. The extensible penis (Fig. 2.6) is located within a hardened (sclerotized) sheath called the aedeagus. The general shape of the aedeagus, the form of its tip, and the shape and position of the sclerotized rings and internal elements are distinctive for many species and give clues not only to species identification

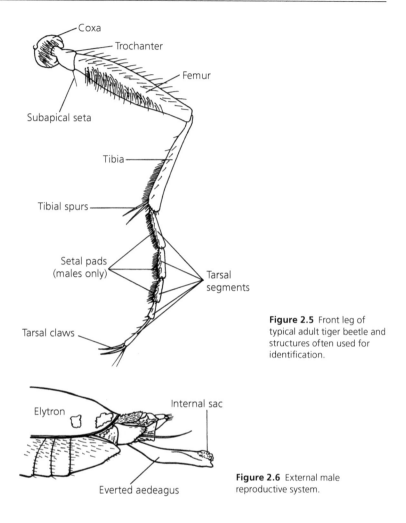

Figure 2.5 Front leg of typical adult tiger beetle and structures often used for identification.

Figure 2.6 External male reproductive system.

but also to phylogenetic relationships among species. When the aedeagus is not extended, it is maintained internally with the right side down. On extension, it rotates 90° clockwise. At the tip of the extended aedeagus, there is a less-sclerotized area (internal sac). This internal sac is turned inside out during copulation to extend beyond the tip of the hard aedeagus, and it delivers sperm or packets of sperm (spermatophores) to the female. Due to the difficulty of observing characters of the aedeagus in a live beetle, they have not been included as part of the identification keys in this field guide.

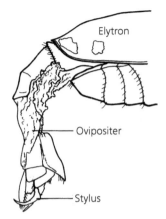

Figure 2.7 External female reproductive system.

The female mating structures (Fig. 2.7) also include some specialized characters. The eighth and ninth abdominal segments are modified to form a telescopic ovipositor, which is used to insert eggs, one at a time, into the substrate. Species differ considerably in the form and shape of the ovipositor, especially the terminal portion (gonapophysis). The differences in these characters have only begun to be studied but offer the potential for many insights into phylogeny, ecology, and behavior.

Larvae

Larval tiger beetles have fewer characters (Fig. 2.8) that distinguish species than do the adults. Important distinctions among larvae are found in the shape and relative size of the inner and median hooks on the back

Figure 2.8 Side view of typical larval tiger beetle and structures often used for identification.

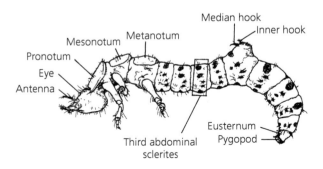

of the fifth abdominal segment. The relative size, number and placement of the simple eyes (stemmata), and relative lengths of the segments of the short antennae are often useful taxonomically. The size and shape and the presence of ridges on the head and dorsal thoracic plates (nota), mouthparts, and terminal abdominal segment (pygopod) can distinguish taxa. Also, sometimes important are subtle differences in the number and position of hair-like setae on the pronotum and throughout the body.

The eyes are the most studied organs of larval tiger beetles. The sedentary larvae have less difficulty detecting prey movement than do the mobile adults. Unlike the grub-like larvae of some insects, which only can achieve a coarse visual pattern with their simple eyes, the eyes of tiger beetle larvae have dense photoreceptors that permit detailed focusing and three-dimensional perception.

Systematics and Taxonomy

Tiger beetle biology, distribution, and natural history are connected by studies of evolutionary relationships (systematics). These systematic studies allow researchers to test specific ideas about which groups of tiger beetles are most closely related to other groups. We call these extended family trees phylogenies. Phylogenies are also important for a practical reason. Instead of randomly arranging tiger beetles in groups, the large number of species conveniently can be organized and categorized on the basis of their relatedness, and these groups can be given names (taxonomy). With an evolutionary basis that can be constantly tested and updated, everyone can communicate efficiently about the organisms by using a reliable and relatively uniform system of names (nomenclature) to discuss a genus, species, or population.

In 1915, Walther Horn developed a phylogeny and classification system for the then-known species of tiger beetles that continues to be used today. After Horn, the most significant advance in our understanding of cicindelid classification was made by Emilie Rivalier in the 1950s to the 1970s. He took advantage of the pattern that in beetles and other insects the shape and structure of male genitalia often provide a unique character system for grouping closely related species. In a series of remarkable articles in the Revue Française d'Entomologie, he meticulously investigated the genitalia of most of the then known species of *Cicindela* on a worldwide basis, with one major biogeographic region in each article. Based on these investigations he designed a classification scheme for all major groups and erected more than 50 groups based on similarities in male genitalia. This system, in which he attempted to group species based on similarity and biogeographic distribution, presumably reflected evolutionary relationships. It found wide acceptance and largely superseded the less explicit scheme of Horn.

For North American taxa of tiger beetles, their names and evolutionary groupings (clades) are generally stable, but as with all scientific endeavors, they are subject to re-interpretation and change as additional information becomes available. The genera of Giant Tiger Beetles (*Amblycheila*) and Night-stalking Tiger Beetles (*Omus*) form distinctive evolutionary groups each considered generally to have been formed from a common ancestor (monophyletic), although controversy surrounds how many species should be recognized within the genus *Omus*. The many species of Metallic Tiger Beetles (*Tetracha*) are now considered to make up a separate genus restricted to the Western Hemisphere. Previously they were considered to be in a subgenus within the worldwide genus *Megacephala*.

Within the evolutionary branch of the Common Tiger Beetles (*Cicindela*), however, the greatest controversy continued. Should Rivalier's subgroups be considered subgenera within the genus *Cicindela* or should the groups actually represent genera of their own? A recent reevaluation of the relationships of North American tiger beetles using nuclear and mitochondrial DNA showed that in some cases Rivalier's subgroups were supported as monophyletic, but in other cases they were not. New groupings of species were erected based on the combined data. Many of these new species groupings were also consistent with shared patterns of body structure (morphology), biogeography, and ecology. In the second edition of this field guide, we follow this most recent interpretation of species groups.

Although modern studies of DNA, molecular clocks, sophisticated statistical programs, and more intense field collecting and observations have provided us a better idea of the phylogeny and classification of tiger beetles, some relationships could change in the future as new information is obtained. With each successive investigation, we come one step closer to understanding the actual evolutionary process that resulted in the tiger beetles we see today.

What Are Species?

Although most biologists agree that plants and animals are made up of discreet groups of similar organisms we call species, how these species are distinguished and which ones are different enough to have separate names are often difficult and sometimes controversial decisions. In an admittedly oversimplified explanation, the evolution of species is an ongoing and dynamic process. At any one time, there are examples of many levels of how different two populations have become in terms of their genes, ecology, behavior, and physical appearance. At what point these differences are sufficient to distinguish them as separate species is usually in the eye of the beholder. There are also many definitions of what constitutes a species, but even experts using the same definition cannot always agree what body structures, behavior or molecules are the best ones on which to base such a decision.

Because naming a species involves so many gray areas, we have decided to use the most widely accepted decisions of tiger beetle experts published in journals. These decisions have been based on distribution data, studies of body structures and now, more and more frequently, on DNA and other molecular clues. However, we also recognize that each of these species decisions is a hypothesis based on the most current information. As more details and analytical techniques become available, some of the named species will be "split" or "lumped" with a new more insightful hypothesis replacing the old one. This is the power of critical thinking and the scientific method, even though it often frustrates those who would prefer final and concrete names for the species they are studying.

What Are Subspecies?

Another aspect of phylogeny involves understanding patterns of differences and similarities among individuals within the same species. Most species show some differences across their geographical ranges. Usually, the larger the range, the more likely there are to be differences in body size, color, and genetic variation associated with geographical locations. Classically, these different identifiable subsets of a species have been called races, morphs, or subspecies. Ernst Mayr called a subspecies "a geographically defined aggregate of local populations which differ taxonomically from other subdivisions of the species."

One unresolved question is whether taxonomic subspecies truly represent evolutionary groupings. Several studies on tiger beetles have demonstrated that these observed differences in color and pattern might simply be the result of local or regional environmental conditions altering their development. Without a heritable basis for a trait, any such variation would not be biologically meaningful. Consequently, some researchers have argued that subspecies identification should be based on genetic traits. To date, the few genetic studies on tiger beetles have shown little to no genetic differentiation in subspecies based on color pattern and size, prompting a re-evaluation of how we define these units.

Furthermore, because different subspecies usually can interbreed freely and exchange genes, their status as a distinctive evolutionary group is vague and subject to considerable change over a relatively short time. Where ranges of different subspecies meet, zones of intergradation are often found, in which colors or other distinguishing characters are intermediate or thoroughly mixed. These zones attest to an extreme plasticity of form and shape.

To formalize what is often an arbitrary geographic distinction, taxonomists may add a third name to the scientific name as convenient shorthand to indicate the identification of a subspecies. The named subspecies, as well, often becomes more interesting to many people, and the patterns of geographical variation more focused. For conservation efforts, subspecies names formalize distinct variations within a species that can make policy decisions more palatable to politicians and legislators. The key to using subspecies, however, is to never forget how imprecise they are and to understand that it is debatable whether they should be used at all or whether other more genetically defined groups should be used in the future. Many researchers studying other groups of organisms have questioned the validity of traditional subspecies or have abandoned their use. Further research is needed to determine whether tiger beetle subspecies should be considered valid in most cases. So it is within this context of more targeted patterns of geographical variation and improved communication tempered by caution for an admittedly vague concept that we include subspecies in the treatment of the tiger beetle fauna of the United States and Canada.

Illustrated Keys to Adult Genera and Species

Diagnostic Characters for Identification

The extensive series of color plates in this field guide includes species, subspecies, and additional variations of United States and Canadian tiger beetles and should enable both amateur and specialist to identify nearly all of the tiger beetles they find, especially when combined with the range maps and descriptive information in the species accounts. However, some species are very similar in color and marking patterns, and in these cases, the use of a taxonomic key that includes more detailed diagnostic characters may be needed. The keys that we use here are based on those developed by Harold Willis and Gary Dunn, but we incorporate some reorganization and modifications that should allow for easier distinctions among the species. We also substitute less technical terminology to make the key more "user-friendly" for the nonspecialist. These characters are described here and illustrated by line drawings throughout the key. In several cases, we also rely on diagnostic geographic distribution to separate otherwise similar species. Be aware, however, that a few species, especially those within the same couplet, may be especially difficult to separate because of the considerable amount of individual variation and overlap in elytral maculations and other characters.

Elytra and Color

The elytra are the first pair of hardened wings that cover the abdomen of beetles, and they are one of the most useful identifying characters. Their color and often distinctive markings (maculations) readily identify many tiger beetles, especially when combined with habitat and geographic range. Elytral colors are usually consistent within a species, but several distinct color forms may occur within a population or species. The color of the underside of the abdomen is usually metallic but in many of the American Tiger Beetles (genus *Cicindelidia*) is distinctly red or red-orange. The most common pattern of lines on the elytra includes three distinct white marks on each elytron (the complete or normal pattern), a front maculation (humeral lunule), a middle maculation (middle band), and a rear (apical lunule) maculation at the tip (Fig. 4.1). In some tiger beetles these maculations are

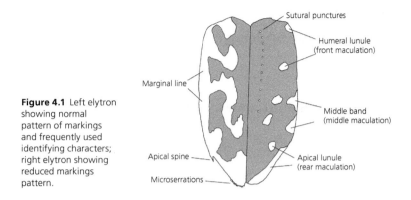

Figure 4.1 Left elytron showing normal pattern of markings and frequently used identifying characters; right elytron showing reduced markings pattern.

reduced to a series of separate dots (Fig. 4.1), some or all may be absent (immaculate), or, at the other extreme, they can be greatly expanded and fused together to form mostly white elytra. The white line running along the outer edge of the elytra (marginal line) may partially or fully connect the three maculations. Within species, subspecies, and populations, there can be considerable variation in the extent of these maculations.

The elytral surface may be shiny, metallic, dull, or greasy in appearance. These differences are caused by texture, primarily surface pits of variable size (seen under ×40 magnification). Deep pits will give a shiny appearance, shallow pits or no pits will make the elytra appear dull. Sometimes larger pits, called punctures, are present, usually in rows running along the length of the middle of the elytra. These pits often reflect a different color than the rest of the elytral surface. Another variation of surface texture is the presence of small raised bumps or granules. Small tooth-like projections (microserrations), seen only at ×40 magnification, are present along the rear margin of the elytra in some species. Distinct extensions or flanges on the outer edge of the elytra are seen in females of a few species.

Setae

The numbers and types of hair-like setae on various parts of the body are useful in distinguishing species and some of the genera. Long, thin setae are often present on the antennae, head, and labrum and at the base of the legs (trochanters). Most species have a row of long sensory setae along the outer end of the first antennal segment (Fig. 4.13) but may or may not have additional setae on other parts of the segment. Groups of long erect setae that originate from medium or large pits are especially common in most species in the Temperate (genus *Cicindela*), American (genus *Cicindelidia*), and American Diminutive (genus *Parvindela*) Tiger Beetles. These setae may be worn off in older specimens, but evidence of their presence can often be deduced from the presence of these peculiar pits. Short, thick, flattened setae on the head and thorax are diagnostic for most species of the Saline (genus *Eunota*), and Ellipsed-winged (genus *Ellipsoptera*) tiger beetles.

Other Characters

The proportional length and the number of teeth (usually one or three) on the labrum are also used in the key. Usually the labrum is wide and short, but in a few species it is longer than it is wide. Body length varies from about 6 mm to over 30 mm and may be used to separate some of the species.

Identification Keys

The first key below is to adults of the genera found in the United States and Canada: the Giant Tiger Beetles (*Amblycheila*), Night-stalking Tiger Beetles (*Omus*), Metallic Tiger Beetles (*Tetracha*), Dromo Tiger Beetles (*Dromochorus*), Ellipsed-winged Tiger Beetles (*Ellipsoptera*), and all the remaining genera often lumped together in the tribe Cicindelini. Separate keys to the species within each of these genera then follow. A few variable species with distinct subspecies key out at more than one place.

Key to the Genera

1a.	Front corners of pronotum lacking a forward projecting lobe (Fig. 4.2B)	2
1b.	Front corners of pronotum prominent with a distinct projecting lobe (Fig. 4.2A)	4

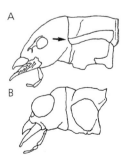

A

B

Figure 4.2 A, Prominent front corner of pronotum of *Tetracha*; **B**, lack of front corner of pronotum of *Cicindela*.

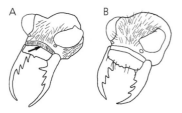

Figure 4.3 A, Clypeus with dense, flattened setae; **B**, clypeus with sparse or no setae.

2a.	Legs covered throughout with fine flattened setae, head proportionally large, body and legs all dark-colored with no maculations	Key IV: Dromo Tiger Beetles (*Dromochorus*)
2b.	Legs with fewer thicker and longer setae, variable in color and maculations	3

3a.	Area of face above labrum (clypeus) densely to sparsely covered with flattened setae (Fig. 4.3A)	Key V: *Eunota togata* and all species of Ellipsed-winged Tiger Beetles (*Ellipsoptera*)
3b.	Area of face above labrum (clypeus) with few long or no setae (Fig. 4.3B)	Key VI: All other genera of tiger beetles in the tribe Cicindelini

4a.	Upperparts pigmented, often with metallic reflections; legs, antennae, and mouth parts unpigmented, tan to yellow in color; elytra often with unpigmented apical maculation	Key III: Metallic Tiger Beetles (*Tetracha*)
4b.	Body, legs, antennae, and mouth parts similarly colored, black to brown to red brown, never metallic reflections, elytra never with maculations	5

5a.	Elytral surface with at least one pleat along each elytron; dorsal surface of pronotum smooth with center line faint to absent; length greater than 20 mm	Key I: Giant Tiger Beetles (*Amblycheila*)
5b.	Elytra never with pleats; length less than 22 mm or if greater than 20 mm the pronotum has a wavy texture and distinct center line	Key II: Night-stalking Tiger Beetles (*Omus*)

Key I: Species of Giant Tiger Beetles (*Amblycheila*)

1a.	Large body size (29–36 mm), found east of Rocky Mountains	2
1b.	Smaller body size (20–28 mm), west of Rocky Mountains	3

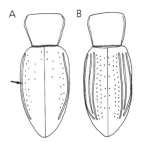

Figure 4.4 A, One pleat running the length of each elytron (Montane Giant Tiger Beetle, *Amblycheila baroni*); **B,** three pleats running the length of each elytron (Mojave Giant Tiger Beetle, *A. schwarzi*).

2a.	Single row of pits between the two inner pleats on elytra; south Texas only	South Texas Giant Tiger Beetle (*A. hoversoni*)
2b.	Entire elytral surface covered with rows of pits; not found in south Texas	Great Plains Giant Tiger Beetle (*A. cylindriformis*)
3a.	Dull black elytra surface, with a single ridge or pleat (rarely two) running the length of each elytron (Fig. 4.4A); central and southeastern Arizona (perhaps extreme west Texas)	Montane Giant Tiger Beetle (*A. baroni*)
3b.	Black or dark brown but shiny and with three pleats running the length of each elytron; not found in southeastern Arizona (Fig. 4.4B)	4
4a.	Entirely black; occurs in intermontane valleys of southern Rocky Mountains of southwest Utah to California	Mojave Giant Tiger Beetle (*A. schwarzi*)
4b.	Head and thorax black, elytra usually dark brown; occurs from southwestern Colorado and southeastern New Mexico to northern Arizona	Plateau Giant Tiger Beetle (*A. picolominii*)

Key II: Species of Night-stalking Tiger Beetles (*Omus*)

1a.	Elytra with many scattered large dimples among smaller punctures, sides of pronotum sharp and lacking setae, dorsal surface with a roughened appearance, often with metallic highlights; 16–20 mm; British Columbia, Washington, and Oregon	Greater Night-stalking Tiger Beetle (*O. dejeani*)
1b.	Elytra with only small indistinct dimples among punctures, dorsal surface dull or shiny, usually smaller in body size	2
2a.	Row of distinct, long, stout, black setae on lateral margins of pronotum, restricted to Warthan Canyon in extreme western Fresno County, California (Fig. 4.5)	Lustrous Night-stalking Tiger Beetle (*O. submetallicus*)

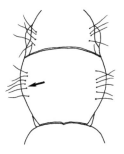

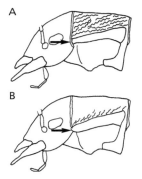

Figure 4.5 Top view of thorax and head of Lustrous Night-stalking Tiger Beetle (*Omus submetallicus*) showing unique, black hairs along sides of pronotum.

Figure 4.6 A, Top of thorax with forward corners slightly turned down (California Night-stalking Tiger Beetle, *Omus californicus*); **B**, forward corners turned distinctly down (Audouin's Night-stalking Tiger Beetle, *O. audouini*).

2b.	Setae absent from upper side of thorax or if present are thin and indistinct	3

3a.	Anterior corners of pronotum visible when viewed from above; 12–23 mm; ranges from montane and coastal regions of central and northern California; on coast as far south as Santa Barbara County in Sierra Nevada (Fig. 4.6A)	California Night-stalking Tiger Beetle (*O. californicus*)
3b.	Anterior corners of pronotum sharply turned down and not visible when viewed from above (Fig. 4.6B)	4

4a.	Pronotal disc with distinct sculpturing, especially laterally; shallow dimples scattered among punctures; 14–18 mm; British Columbia to northwest California	Audouin's Night-stalking Tiger Beetle (*O. audouini*)
4b.	Pronotal disc with distinct sculpturing, less distinct laterally, dimples obscured by deep pits; 14–16 mm, southern Oregon in area of Mt. Ashland	Mount Ashland Night-stalking Tiger Beetle (*O. cazieri*)

Key III: Species of Metallic Tiger Beetles (*Tetracha*)

1a.	Elytra dark, oily green, lacking light-colored maculations	Virginia Metallic Tiger Beetle (*T. virginica*)
1b.	Elytra bright maroon or dark green with light-colored maculations at rear tip	2

2a.	Rear maculations not expanded at their front ends	Upland Metallic Tiger Beetle (*T. impressa*)
2b.	Rear maculations crescent-shaped and distinctly expanded at their front ends	3

| **3a.** | Upper surface dark green | Florida Metallic Tiger Beetle (*T. floridana*) |
| **3b.** | Upper surface iridescent maroon and green | Carolina Metallic Tiger Beetle (*T. carolina*) |

Key IV: Species of Dromo Tiger Beetles (*Dromochorus*)

| **1a.** | Elytral surface with distinct sutural punctures running lengthwise along middle of each elytron (sometimes with metallic reflections) | 2 |
| **1b.** | Elytral surface lacking distinct sutural punctures | 4 |

| **2a.** | Elytral sutural punctures with metallic green, blue, or gold reflections. Textured swirls present on central part of elytra (Fig. 4.7) | 3 |
| **2b.** | Elytral sutural punctures never with metallic reflections. Color dark gray to black. Elytra dull textured or only slightly velvety. Elytral swirls usually not present. Oklahoma and panhandle of Texas, south to east central Texas | Loamy-ground Tiger Beetle (*D. belfragei*) |

| **3a.** | Elytra with shallow sutural punctures with metallic green reflections. Ground color dark olive-brown, occasionally with extensive green on the head, pronotum, and elytra. Male labrum entirely white or pale. Louisiana to eastern Texas | Cajun Tiger Beetle (*D. pilatei*) |

Figure 4.7 Elytral swirls of dark color.

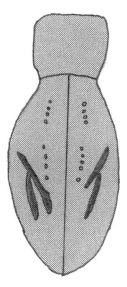

Figure 4.8 Black labrum of male with white central spot.

3b.	Elytra with deep sutural punctures with metallic green or gold reflections. Ground color charcoal to jet-black. Male labrum black with pale center (Fig. 4.8). "Hill Country" region of central Texas	Juniper Grove Tiger Beetle (*D. knisleyi*)
4a.	Elytra strongly velvety or frosty in texture. Violet or blue reflections may be present on elytra	5
4b.	Elytral texture dull or only slight velvety. Ground color dark gray to jet-black. Never with violet or blue reflections on elytra. Oklahoma and panhandle of Texas, south to east central Texas	Loamy-ground Tiger Beetle (*D. belfragei*)
5a.	Maxillary palps yellow or pale with contrasting dark terminal segment (Fig 4.9A). Elytra velvety in texture, always with a strong violet or blue sheen throughout. Kansas and western Missouri, south to central Texas	Frosted Tiger Beetle (*D. pruininus*)
5b.	Maxillary palps dark brown to black (Fig 4.9B), sometimes with metallic green and violet reflections	6
6a.	Elytra finely velvety black, with strong violet or blue reflections, especially along margins. Body very narrow. Male labrum entirely dark. extreme south Texas, south of Corpus Christi	Velvet Tiger Beetle (*D. velutinigrens*)

Figure 4.9 A, Pale maxillary palps with contrasting dark tips; **B,** maxillary palps dark throughout.

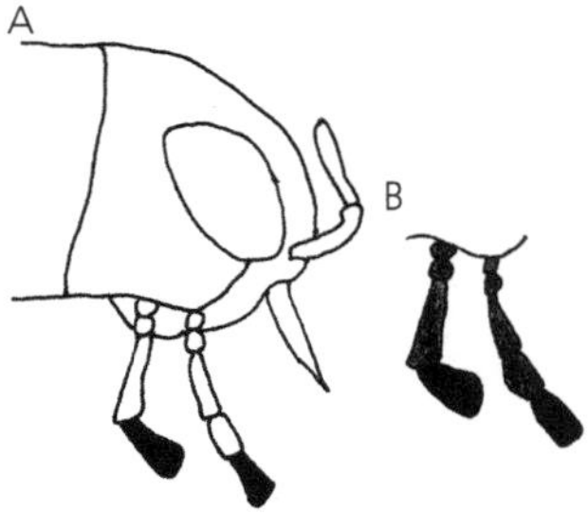

6b.	Elytra velvety or frosty in texture, usually ashy-gray to black, occasionally with faint violet or blue reflections. Male labrum with pale center spot. (Fig. 4.9)	7

7a.	Elytra frosty in texture. Color ashy-gray to black, occasionally with blue reflections. South Texas in mesquite chaparral forest. Known only from Bexar, Frio, and Atascosa Counties	Pygmy Dromo Tiger Beetle (*D. minimus*)
7b.	Elytra finely velvety black, rarely with blue reflections. South Texas to Mexico in mesquite chaparral forest. Known from Dimmit, LaSalle, and Webb Counties in Texas, and the state of Tamaulipas, Mexico	Chaparral Tiger Beetle (*D. chaparrelensis*)
7c.	Elytra finely velvety black, rarely with blue reflections. Found in coastal prairie habitat near Gulf of Mexico	Gulf Prairie Tiger Beetle (*D. welderensis*)

Key V: Species of Ellipsed-winged Tiger Beetles (*Ellipsoptera*) and the White-cloaked Tiger Beetle (*Eunota togata*).

1a.	Underside of thorax (prosternum) with dense flattened setae (Fig. 4.10)	2
1b.	Underside of thorax (prosternum) lacking setae	6

2a.	Elytral surface dull and lacking pits; maculations diffuse and broadly expanded	3
2b.	Elytral surface dull to shiny with many deep pits; maculations variable	4

3a.	Labrum with few (<10) or no flattened setae (not including marginal row); inner margins of white maculations ragged; north Florida to Virginia	Whitish Tiger Beetle (*Ellipsoptera gratiosa*)
3b.	Labrum densely covered with many (>20) flattened setae; inner margins of white maculations more distinct; restricted to peninsular Florida and southeastern Georgia	Moustached Tiger Beetle (*Ellipsoptera hirtilabris*)

4a.	Sides of pronotum distinctly convex; middle maculation not prominent; broad white band along outer edge of elytra or elytra entirely white	White-cloaked Tiger Beetle (*Eunota togata*)

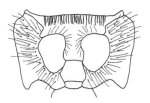

Figure 4.10 Dense, flattened setae on underside of thorax.

4b.	Sides of pronotum straight or slightly curved; long middle maculation	5

5a.	Rear end of female elytra bent downward at tips, apical spine slightly retracted (Fig. 4.11C); right mandible of male with prominent tooth below; Atlantic coast and south Florida Gulf coast (Fig. 4.11A)	Margined Tiger Beetle (*Ellipsoptera marginata*)
5b.	Rear end of female elytra not bent downward at tips, apical spine much retracted (Fig. 4.11D); right mandible of male with bump or no tooth below (Fig. 4.11B); Gulf coast, Florida to Texas	Coastal Tiger Beetle (*Ellipsoptera hamata*)

6a.	Legs lacking pigment (pale tan); distinct cluster of setae on fourth antennal segment	Ghost Tiger Beetle (*Ellipsoptera lepida*)
6b.	Legs pigmented dark	7

7a.	Front maculation with a prominent dot at front end of each elytron forming a "G" or inverted "G" (Fig. 4.12A, B)	8
7b.	Front maculation lacking a basal dot at front end of each elytron forming a "J" or inverted "J" (Fig. 4.12C)	12

8a.	Medium sized (>11 mm); middle maculation long	9
8b.	Small (<11 mm); middle maculation very short, broad basally and tapering at its inner end	White-sand Tiger Beetle (*Ellipsoptera wapleri*)

9a.	Middle maculation "S"-shaped, with rear portion turning forward; restricted to coastal plain of southeastern United States (Fig. 4.12A)	Sandbar Tiger Beetle (*Ellipsoptera blanda*)
9b.	Middle maculation with rear portion usually not turning forward (Fig. 4.12B)	10

Figure 4.11 A, Male right mandible with distinct tooth on underside (Margined Tiger Beetle, *Ellipsoptera marginata*); **B**, male right mandible lacking distinct tooth on underside (Coastal Tiger Beetle, *E. hamata*); **C**, rear end of female elytron bent downward with apical spine near tip; **D**, rear end of female elytron unbent and with apical spine farther forward.

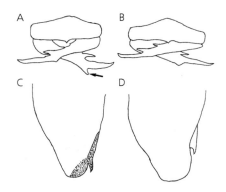

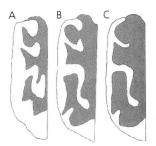

Figure 4.12 A, Middle maculation "S"-shaped and front maculation "G"-shaped; **B**, maculation not "S"-shaped and front maculation "G"-shaped; **C**, maculation not "S"-shaped and front maculation "J"-shaped.

10a.	East of Appalachian Mountains, Chesapeake Bay to New Hampshire	Puritan Tiger Beetle (*Ellipsoptera puritana*)
10b.	West or south of Appalachian Mountains	11
11a.	Elytra dull, pits smaller and more shallow; tip ends of the female elytra pointed, without a small notch between them; rear end of middle maculation recurved	Sandy Stream Tiger Beetle (*Ellipsoptera macra*)
11b.	Elytra shiny, pits larger and deeper; tips of the elytra of female rounded with a small notch between them; rear end of middle maculation globular	Coppery Tiger Beetle (*Ellipsoptera cuprascens*)
12a.	First antennal segment with few to many setae in addition to sensory setae (Fig 4.13A); middle maculation usually hooked at the rear and only slightly "S"-shaped	Nevada Tiger Beetle (*Ellipsoptera nevadica*)
12b.	First antennal segment with only long sensory setae and no small setae (Fig 4.13B); middle maculation usually not hooked at the rear and not "S"-shaped	13

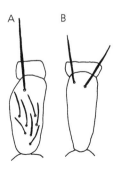

Figure 4.13 A, First antennal segment with small setae in addition to long sensory setae; **B**, first antennal segment with only two long sensory setae.

| 13a. | Sides of thorax coppery; elytral surface dull brown to reddish-brown | Rio Grande Tiger Beetle (*Ellipsoptera sperata*) |
| 13b. | Sides of thorax blue-green to blue-purple (may be coppery at upper edge); elytral surface shiny green or reddish-brown | Aridland Tiger Beetle (*Ellipsoptera marutha*) |

Key VI: Species of Other Genera Within the Tribe Cicindelini

| 1a. | Front trochanters with one (rarely two) subapical setae, middle trochanters with or without such setae (Fig. 4.14A) | 16 |
| 1b. | Front trochanters lacking subapical setae, middle trochanters also lacking such setae (Fig. 4.14B) | 2 |

| 2a. | Femora of hind legs long, extending for more than one-third of their length beyond end of body | 3 |
| 2b. | Femora of hind legs short, typically not extending more than one-third of their length beyond end of body | 4 |

| 3a. | Underside of thorax (prosternum) with dense flattened setae (Fig. 4.10) | Eastern Beach Tiger Beetle (*Habroscelimorpha dorsalis*) |
| 3b. | Underside of thorax (prosternum) lacking setae; known only from old records on far south Texas coast | Lime-headed Tiger Beetle (*Opilidia chlorocephala*) |

4a.	Abdominal segments green with some brown (Louisiana to west Texas) or last few segments all brown (New Mexico and Arizona); maculations consisting of a series of four nearly equally separated dots	Ocellated Tiger Beetle (*Cicindelidia ocellata*)
4b.	Most abdominal segments green or green with brown; only last few abdominal segments entirely orange or brown; maculations consist of thin lines and spots; restricted to Chiricahua mountains of southeastern Arizona	Melissa's Tiger Beetle (*Cicindelidia melissa*)
4c.	Abdominal segments orange, orange-red, or rarely brown, maculations not as above	5

Figure 4.14 A, Trochanter of middle leg with setae; **B**, trochanter with no setae.

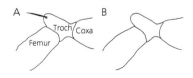

5a.	Elytra green to green-brown with a wide, light-colored border along the outer edge	Cobblestone Tiger Beetle (*Cicindelidia marginipennis*)
5b.	Elytra lacking above combination of characters, usually black to brown, rarely green	6

6a.	Elytra dull black with wide maculations, including a wide straight middle maculation that angles back away from the outer edge	.. Schaupp's Tiger Beetle (*Cicindelidia schauppii*)
6b.	Elytra dull or shiny, black to brown or sometimes blue, lacking wide or angled maculations	7

7a.	Elytra with three orangish maculations; accidental on the border of southern Arizona and New Mexico	Red-lined Tiger Beetle (*Cicindelidia fera*)
7b	Maculations whitish and thin or reduced	8

8a.	Labrum with four setae (Fig. 4.15A,C)	9
8b.	Labrum with six setae (Fig. 4.15E)	10

9a.	Sides of underparts of thorax and abdomen sparsely covered with flattened setae; pronotum with some setae; northern Florida to New York (Fig. 4.15B)	Eastern Pinebarrens Tiger Beetle (*Cicindelidia abdominalis*)
9b.	No setae on underparts of abdomen or on pronotum; restricted to Polk and Highlands Counties in central Florida (Fig. 4.15D)	Highlands Tiger Beetle (*Cicindelidia highlandensis*)

10a.	Elytral surface with deep pits (Fig 4.8F) and shiny surface; many setae on sides of abdomen and thorax; small (<9 mm); restricted to Florida peninsula and/or extreme southeast Georgia	11
10b.	Elytral surface variable, larger in size (>8 mm); not found in Florida peninsula	12

11a	Color above darkish-green, Miami area only	Miami Tiger Beetle (*Cicindelidia floridana*)

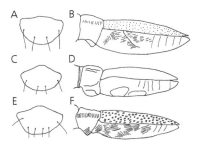

Figure 4.15 A and **C**, Labra with four setae; **E**, labrum with six setae; **B**, sides of thorax and abdomen with sparse setae (Eastern Pine barrens Tiger beetle, *Cicindelidia abdominalis*); **D**, sides of thorax and abdomen with no setae (Highlands Tiger Beetle, *C. highlandensis*); **F**, sides of thorax and abdomen with dense setae and upper surface of elytra with deep pits (Scabrous Tiger Beetle, *C. scabrosa*).

11b	Color above blackish, Florida peninsula north of Miami to southeast Georgia	**Scabrous Tiger Beetle (*Cicindelidia scabrosa*)**
12a.	Elytra with two parallel rows of shallow green pits (punctures) along inner edge of each elytron, restricted to Rio Grande Valley of south Texas	**Cazier's Tiger Beetle (*Cicindelidia cazieri*)**
12b.	Elytra lacking shallow green pits (punctures); wider ranging	13
13a.	Elytral surface shiny, variable colors, maculations absent, reduced, or limited to outer portion of elytra	**Limestone Tiger Beetle (*Cicindelidia politula*)**
13b.	Elytral surface not shiny, maculations variable	14
14a	Maculations reduced to thin lines and spots; south Texas to Massachusetts	**Eastern Red-bellied Tiger Beetle (*Cicindelidia rufiventris*)**
14b.	Usually with distinct middle maculation; from west Texas.	15
15a.	Middle maculation slightly constricted in middle (Fig. 4.16A); elytra of female and male expanded in middle	**Wetsalts Tiger Beetle (*Cicindelidia hemorrhagica*)**
15b.	Middle maculation strongly constricted in middle (Fig. 4.16B); elytra of female not expanded; elytra of male widened in rear one-fourth	**Western Red-bellied Tiger Beetle (*Cicindelidia sedecimpunctata*)**
16a.	Very small (<10 mm) red, red-brown to brown; trochanters of middle legs lacking long setae (Fig. 4.14B); restricted to southwestern United States	17
16b	Size variable but if small not found in southwestern United States; trochanters of middle legs with one or rarely two long setae (Fig. 4.14A)	19

Figure 4.16 A, Elytron with middle maculation slightly constricted (Wetsalts Tiger Beetle, *Cicindelidia hemorrhagica*); **B**, elytron with middle maculation strongly constricted (Western Red-bellied Tiger beetle, *C. sedecimpunctata*).

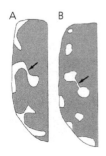

17a.	Metallic red; maculations consist of a longitudinal band running length of elytra near outer edge	White-striped Tiger Beetle (*Parvindela lemniscata*)
17b.	Brown or red brown; maculations consist of dots or short stripes	18

18a.	Elytra dark brown with distinct green punctures and pits; underside of abdomen with sparse flattened setae along outer edge	Pygmy Tiger Beetle (*Brasiella viridisticta*)
18b.	Elytra red to coppery red, lacking distinct punctures and pits; lacking dense flattened setae on underside of abdomen	Sonoran Tiger Beetle (*Brasiella wickhami*)

19a.	Area of forehead between eyes (frons) lacking setae (except for one or two along inner edge of each eye) or with a cluster of flattened setae above the antennae	20
19b.	Area of forehead between eyes (frons) with erect setae	51

20a.	Area of forehead between eyes (frons) with small clusters of flattened setae above antennae (Fig. 4.17A); restricted to Gulf Coast of Texas, rarely Louisiana or Mississippi	Gulfshore Tiger Beetle (*Eunota pamphila*)
20b.	21Area of forehead between eyes (frons) lacking setae (except for one or two along inner edge of eye) (Fig. 4.17B)	21

21a.	Small (>9 mm), dark brown beetles; prothorax cylindrical with nearly straight sides	22
21b.	Lacking above combination of characters	23

Figure 4.17 A, Frons with clusters of setae near each eye; **B**, frons with no clusters of setae.

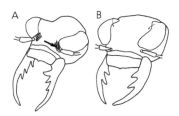

Figure 4.18 A, Elytron of male Swift Tiger Beetle (*Parvindela celeripes*) with posterior portion not as distinctly expanded but maculations reduced; **B**, elytron of female Swift Tiger Beetle with distinctly expanded posterior portions and reduced maculations.

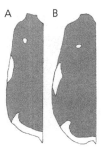

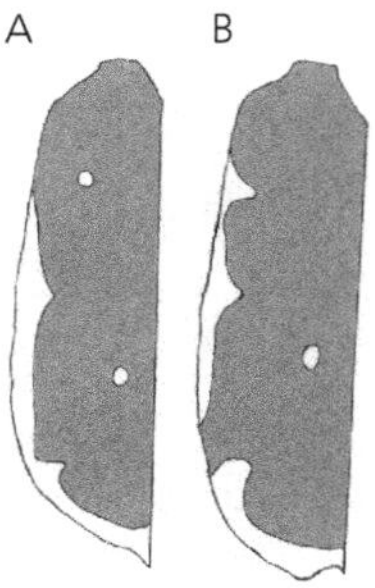

A   B

Figure 4.19 A, Elytron of male Ant-like Tiger Beetle (*Parvindela cursitans*) with posterior portion indistinctly expanded; **B,** elytron of female Ant-like Tiger Beetle with indistinctly expanded posterior portions and continuous maculation along outer edge.

22a.	Elytra notably expanded in posterior one-half; maculations usually reduced, no incomplete whitish line or no line at all along the outer edge (Fig. 4.18)	Swift Tiger Beetle (*Parvindela celeripes*)
22b.	Elytra not notably expanded in posterior one-half; usually with complete whitish line along outer edge (Fig. 4.19)	Ant-like Tiger Beetle (*Parvindela cursitans*)
23a.	Side of thorax (proepisternum) with setae (occasionally only a few near base of leg)	24
23b.	Side of thorax (proepisternum) lacking setae	25
24a.	Labrum medium or short	26
24b.	Labrum longer than wide	34
25a.	Large (>12 mm), brown with single white spot at middle of outer edge of elytra; labrum long, not found in southwestern United States	One-spotted Tiger Beetle (*Apterodela unipunctata*)
25b.	Small (<12 mm) and green with a white band along much of outer edge of elytra; labrum short; restricted to southwestern United States	Grass-runner Tiger Beetle (*Parvindela debilis*)
26a.	Rear edge of elytra lacking small sawtooth edge (microserrations) (visible at ×50 magnification)	27
26b.	Rear edge of elytra with small sawtooth edge (microserrations) along rear end	35
27a.	With cluster of long setae near front of eyes and often additional setae on forehead (Fig. 4.20A); east of Rocky Mountains	28
27b.	With two long setae near front of eyes (Fig. 4.20B)	29
28a.	Elytral surface dull; middle tooth of labrum shorter than teeth on either side of it; male with a black labrum with two white spots; active primarily in the fall	Autumn Tiger Beetle (*Cicindela nigrior*)
28b.	Elytral surface shiny; middle tooth of labrum not shorter than other two; male with all white labrum; active both fall and spring	Festive Tiger Beetle (*Cicindela scutellaris*)

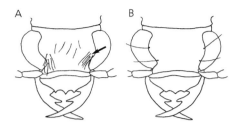

Figure 4.20 A, Cluster of setae in front of eye; **B**, two setae on top and near front of eyes.

29a.	First antennal segment with one (rarely two) seta	30
29b.	First antennal segment with two to four setae	Black Sky Tiger Beetle (*Cicindelidia nigrocoerulea*)
30a.	Sides of abdomen with no setae	Horn's Tiger Beetle (*Cicindelidia hornii*)
30b.	Sides of abdomen with few to sparse flattened setae	31
31a.	Pronotum lacking setae; elytra shiny with distinct pits along inner edges; Florida to South Carolina	Elusive Tiger Beetle (*Eunota striga*)
31b.	Pronotum with at least some setae along outer edge; elytra rarely shiny; with shallow or no pits; central to western states and provinces	32
32a.	Large, robust species (>15 mm); rear half of elytra lacking punctures or pits	Large Grassland Tiger Beetle (*Cicindelidia obsoleta*)
32b.	Small species (<15 mm); rear half of elytra with pits throughout	33
33a.	Historically restricted to western slope of Sierra Nevada Mountains in California and adjacent salt flats in San Joaquin Valley	Meadow Tiger Beetle (*Parvindela lunalonga*)
33b.	Widely distributed from Minnesota and Manitoba west to Washington, Oregon, and east of the Sierra Nevada in extreme northeastern California south to central Arizona	Variable Tiger Beetle (*Parvindela terricola*)
34a.	Underside usually black with purple or blue-green reflections; elytral surface with shallow to deep pits, black, shiny between pits; male labrum white, female labrum partially or completely dark	("nebraskana" variant of Long-lipped Tiger Beetle (*Cicindela longilabris*)
34b.	Underside usually blue-green or green with some purple and sometimes blue or coppery; elytra usually with bumps or punctures, dull or slightly shiny, range from black to bronze to green to blue; usually with some evidence of maculations; labrum of both male and female white	Long-lipped Tiger Beetle (*Cicindela longilabris*)

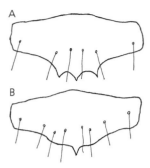

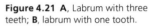

Figure 4.21 A, Labrum with three teeth; **B**, labrum with one tooth.

Figure 4.22 A, Typical middle maculation of Dispirited Tiger Beetle (*Cicindela depressula*) with front portion of middle maculation not concave; **B**, typical middle maculation of Western Tiger Beetle (*C. oregona*) with front portion of middle maculation concave.

35a.	First antennal segment with three or four long setae	36
35b.	First antennal segment with one long seta	41
36a.	Labrum with one tooth; occurs west of central Great Plains (Fig. 4.21B)	37
36b.	Labrum with three teeth; occurs east of central Great Plains (Fig. 4.21A)	39
37a.	Elytra with whitish line along outer edge; lower forehead (frons) with one pair of setae, now restricted to Salmon River of Idaho	Columbia River Tiger Beetle (*Cicindela columbica*)
37b.	Elytra lacking whitish line along outer edge; lower forehead (frons) with cluster of setae	38
38a.	Two or three (rarely four) setae above eyes; central, front portion of middle maculation not concave (if present) (Fig. 4.22A)	Dispirited Tiger Beetle (*Cicindela depressula*)
38b.	Clusters of 8 to 11 setae above eyes; central, front portion of middle maculation concave (Fig. 4.22B)	Western Tiger Beetle (*Cicindela oregona*)
39a.	Elytral surface dull; middle maculation usually complete; outer edge of abdomen with sparse flat setae	Northern Barrens Tiger Beetle (*Cicindela patruela*)
39b.	Elytral surface shiny, with shallow to deep pits; outer edge of abdomen lacking setae; middle maculation usually broken into dots or absent	40
40a.	Green to olive green above; restricted to southeastern Manitoba northern Minnesota and southwestern Ontario; (13–15 mm)	Laurentian Tiger Beetle (*Cicindela denikei*)

40b.	Bright metallic green (rarely purple) above; throughout most of eastern United States, southeastern Canada; (10–14 mm)	Six-spotted Tiger Beetle (*Cicindela sexguttata*)
41a.	Maculations incomplete, broken into dots or absent	42
41b.	Maculations complete, often fused	44
42a.	Labrum with 8–10 setae	S-banded Tiger Beetle (*Cicindelidia trifasciata*)
42b.	Labrum with fewer than eight setae	43
43a.	Labrum with one tooth (Fig 4.21B)	Punctured Tiger Beetle (*Cicindelidia punctulata*)
43b.	Labrum with three teeth (Fig. 4.21A)	Saltmarsh Tiger Beetle (*Eunota severa*)
44a.	Middle maculation short, diagonal, or represented by a bulge	45
44b.	Middle maculation "normal," very long and slender, or diagonal and long	48
45a.	Cheeks (genae) with sparse flattened setae (Fig. 4.23)	Riparian Tiger Beetle (*Eunota praetextata*)
45b.	Cheeks (genae) lacking setae	46
46a.	Middle maculation a sharp or rounded bulge, only slightly lengthened toward the rear (Fig. 4.24)	Cream-edged Tiger Beetle (*Eunota circumpicta*)

Figure 4.23 Flattened hair-like setae on sides of face (genae) (Riparian Tiger Beetle, *Eunota praetextata*).

Figure 4.24 Middle maculation reduced to rounded bulge with little (**A**) or no (**B**) extension to the rear (Cream-edged Tiger Beetle, *Eunota circumpicta*).

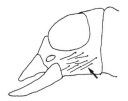

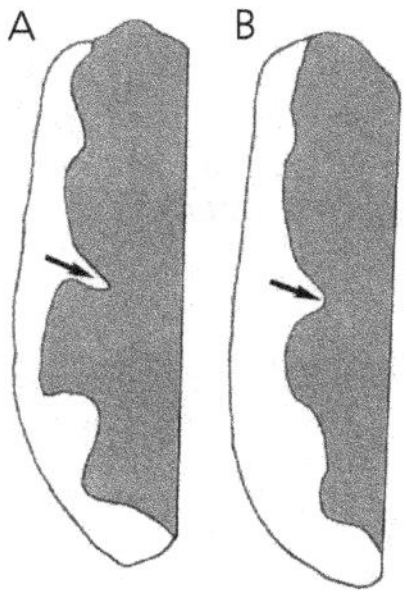

Figure 4.25 A, Typical maculation pattern of the California Tiger Beetle (*Eunota californica*), in which the dark wedge between the middle and rear maculations extends forward less or about the same distance as rearward; **B–D,** range of maculation patterns for the Glittering Tiger Beetle (*E. fulgoris*), in which the dark wedge between the middle and rear maculations, if present, extends farther forward that rearward.

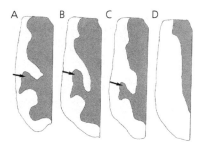

46b.	Middle maculation more lengthened toward the rear or fused (Fig. 4.25)	47
47a.	Found in New Mexico, west Texas, and eastern Arizona (Fig. 4.25B-D)	Glittering Tiger Beetle (*Eunota fulgoris*)
47b.	Found in southeastern California (Fig. 4.22A)	California Tiger Beetle (*Eunota californica*)
48a.	Underside of thorax (prosternum) with a few erect setae, rear maculation extends far forward paralleling outer edge of elytron (occasionally reduced); restricted to southern California coast	Western Tidal Flat Tiger Beetle (*Eunota gabbii*)
48b.	Underside of thorax (prosternum) lacking setae, rear maculation "normal"	49
49a.	Middle maculation elaborate and "S"-shaped (Fig. 4.26A)	S-banded Tiger Beetle (*Cicindelidia trifasciata*)
49b.	Middle maculation "normal" (Fig. 4.26B)	50
50a.	Rear tip of abdomen red; restricted to Florida Keys	Olive Tiger Beetle (*Microthylax olivacea*)
50b.	Rear tip of abdomen not red; occurs in southwestern United States (Fig. 4.26B)	Thin-lined Tiger Beetle (*Cicindelidia tenuisignata*)
51a.	Cheeks (genae) with numerous (Fig. 4.23) to a few setae at bottom edge	52
51b.	Cheeks (genae) lacking setae	76
52a.	Labrum with one tooth or no teeth (Fig. 4.21B)	53
52b.	Labrum with three teeth (Fig. 4.21A)	56
53a.	First antennal segment with four or five long setae (Fig. 4.27B); sand dune habitat	54

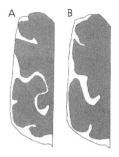

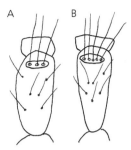

Figure 4.26 A, Strongly "S"-shaped middle maculation of S-banded Tiger Beetle (*Cicindelidia trifasciata*); **B**, more normal-shaped middle maculation of the Thin-lined Tiger Beetle (*C. tenuisignata*).

Figure 4.27 A, First antennal segment with three long setae; **B**, first antennal segment with four or five long setae.

53b.	First antennal segment with three long setae (Fig. 4.27A); variable habitats but usually riparian	55

54a.	Restricted to coastal sand dunes from northern California to Washington	Pacific Coast Tiger Beetle (*Cicindela bellissima*)
54b.	Restricted to Coral Pink Sand Dunes in southwest Utah	Coral Pink Sand Dune Tiger Beetle (*Cicindela albissima*)
54c.	Restricted to Great Sand Dunes in southern Colorado	Colorado Dune Tiger Beetle (*Cicindela theatina)*
54d.	Restricted to Bruneau dune area of southwest Idaho and shows distinct tooth on mandible projecting upward (Fig. 4.28)	Bruneau Dune Tiger Beetle (*Cicindela waynei*)
54e.	Restricted to St. Anthony dune area of central and eastern Idaho	St. Anthony Dune Tiger Beetle (*Cicindela arenicola*)
54f.	Found in Great Plains of United States and Canada with isolated colonies in Labrador and northwest Alaska	Sandy Tiger Beetle (*Cicindela limbata*)

Figure 4.28 Distinctive tooth on mandible of the Bruneau Dune Tiger beetle (*Cicindela waynei*) projecting upward.

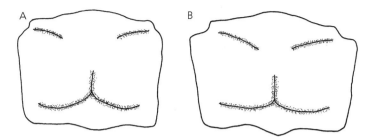

Figure 4.29 A, Cylindrical or symmetrical thorax shape (Bronzed Tiger Beetle, *Cicindela repanda*); **B**, trapezoidal thorax shape (Twelve-spotted Tiger Beetle, *C. duodecimguttata*).

55a.	Front maculation usually complete and connected to or only slightly separated from white line along outer edge of elytra; pronotum narrow with front about the same width as back (proportion: four units long to five units wide)(Fig. 4.29A)	Bronzed Tiger Beetle (*Cicindela repanda*)
55b.	Front maculation usually broken and widely separated from outer edge of elytra and white line along this outer edge usually absent); pronotum trapezoidal in shape with front wider (proportion: four units long to six units wide) (Fig. 4.29B) than back	Twelve-spotted Tiger Beetle (*Cicindela duodecimguttata*)
56a.	Front maculation complete, or so broad it becomes obscured by heavy side extension (front maculation extends almost to the middle of the elytron)	57
56b.	Front maculation absent, broken into dots, or, if complete, projects only slightly toward middle	60
57a.	Maculations connected along outer edge of elytra	58
57b.	Maculations not connected along outer edge of elytra	Oblique-lined Tiger Beetle (*Cicindela tranquebarica*)
58a.	Front maculation separate from middle maculation	Oblique-lined Tiger Beetle (*Cicindela tranquebarica diffracta*)
58b.	Front maculation connected to middle maculation along outer edge of elytra	59
59a.	Labrum long (width/length ratio <1.65); body usually <15 mm long and proportionately slender	Blowout Tiger Beetle (*Cicindela lengi*)
59b.	Labrum medium (width/length ratio >1.65); body usually >15 mm long and proportionately stout	Big Sand Tiger Beetle (*Cicindela formosa*)

60a.	Middle maculation long, reaching back nearly to rear maculation; maculations not connected along outer edge of elytra	61
60b.	Middle maculation usually short or absent; maculations often connected along outer edge of elytra	63
61a.	Front maculation long, straight, and angling away from the outer edge of the elytra (Fig. 4.30A) or absent	Oblique-lined Tiger Beetle (*Cicindela tranquebarica*)
61b.	Front maculation not as above	62
62a.	Elytra reddish-brown to dark greenish-brown; all maculations thin but usually complete; eastern species	Appalachian Tiger Beetle (*Cicindela ancocisconensis*)
62b.	Elytra purplish, red, green, or black; maculations variable but usually reduced and thick; front maculation short or absent, often two dots (Fig. 4.30B) and projects only slightly toward the inner edge of the elytron; western species	Badlands Tiger Beetle (*Cicindela decemnotata*)
63a.	Maculations absent except for small spot at rear tip of elytra; restricted to southeastern New Mexico and far west Texas	Big Sand Tiger Beetle subspecies (*Cicindela formosa rutilovirescens*)
63b.	Maculations apparent; not present in southeastern New Mexico or far west Texas	64
64a.	Head and pronotum much differently colored than elytra	65
64b.	Head, pronotum, and elytra similarly colored (although there may be contrasting colors running along outer edge of elytra)	66
65a.	Head and pronotum copper, elytra blue-green or green	Cow Path Tiger Beetle (*Cicindela purpurea*)
65b.	Head and pronotum green to blue or blue-purple, elytra red to green-red	Splendid Tiger Beetle (*Cicindela splendida*)

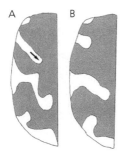

Figure 4.30 A, Long, straight front maculation angled away from outer edge of elytron (Oblique-lined Tiger Beetle, *Cicindela tranquebarica*); **B**, front maculation short and reduced to two dots (Badlands Tiger Beetle, *C. decemnotata*).

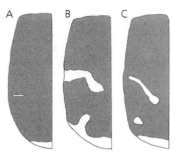

Figure 4.31 A, Middle maculation a transverse dash; **B,** middle maculation with sharp bend; **C,** middle maculation with rounded bend.

66a.	Middle maculation absent or reduced to a transverse dash (Fig. 4.31A)	67
66b.	Middle maculation complete or consists of an angled, bent line (Fig. 4.31B)	71
67a.	Sagebrush and alpine areas of California, Nevada, Oregon, and Idaho; labrum of female black	Alpine Tiger Beetle (*Cicindela plutonica*)
67b.	Restricted to Death Valley area of eastern California; labrum of female whitish	Oblique-lined Tiger Beetle (*Cicindela tranquebarica arida*)
67c.	East of Rocky Mountains	68
68a.	Elytra black	Cow Path Tiger Beetle (*Cicindela purpurea*)
68b.	Elytra not black	69
69a.	Elytra green or blue-green	70
69b.	Elytra reddish or dark red	Common Claybank Tiger Beetle (*Cicindela limbalis*)
70a.	Occurs in northwest Louisiana, southwest Arkansas, and northeast Texas	Splendid Tiger beetle (*Cicindela splendida* "ludoviciana")
70b.	Occurs in the Great Plains	Green Claybank Tiger Beetle (*Cicindela denverensis*)
71a.	Maculations not connected along outer edge of elytra	72
71b.	Maculations connected along outer edge of elytra or middle maculation broadly expanded along outer edge of elytra	75

72a.	Restricted to coastal Santa Cruz County, California; front maculation reduced to one or two dots	Ohlone Tiger Beetle (*Cicindela ohlone*)
72b.	Not found in Santa Cruz County, California	73
73a.	Middle maculation with only a slight or rounded bend, directed posteriorly toward midline (Fig. 4.31C); front maculation absent	Cow Path Tiger Beetle (*Cicindela purpurea*)
73b.	Middle maculation with sharp bend (Fig. 4.31B); front maculation complete or broken into dots	74
74a.	Head, pronotum, and elytra bright metallic green or blue; middle maculation usually reduced but may be complete	Green Claybank Tiger Beetle (*Cicindela denverensis*)
74b.	Head, pronotum, and elytra brick red or cuprous	Common Claybank Tiger Beetle (*Cicindela limbalis*)
75a.	Elytra dark green to dark red-green to black	Cow Path Tiger Beetle (*Cicindela purpurea cimarrona*)
75b.	Elytra bright red, bright red-green, or bright green	Common Claybank Tiger Beetle (*Cicindela limbalis*)
76a.	Underside of thorax (prosternum) with sparse erect setae; maculations complete and heavy; upper surface blackish; restricted to beaches of coastal southern California	Western Beach Tiger Beetle (*Cicindela latesignata*)
76b.	Lacking these characters	77
77a.	Elytral surface smooth and lacking pits or bumps	Festive Tiger Beetle (*Cicindela scutellaris*)
77b.	Elytral surface uneven with small bumps, pits, or granules	78
78a.	First antennal segment with no to two long, erect setae in addition to small sensory setae	79
78b.	First antennal segment with more than two long, erect setae in addition to small sensory setae	81
79a.	Front maculation present, long and angling away from outer edges of elytra	Pacific Coast Tiger Beetle (*Cicindela bellissima*)
79b.	Front maculation absent or, if present, not angling away from outer edges of elytra or obvious and projecting toward the front	80
80a.	With many long setae on lower forehead (frons)	Hairy-necked Tiger Beetle (*Cicindela hirticollis*)

43

80b.	No setae on lower forehead (frons) but with clusters of two to four setae above eyes	Dispirited Tiger Beetle (*Cicindela depressula*)
81a.	Labrum long (width/length ratio <1.9)	Dark Saltflat Tiger Beetle (*Cicindela parowana*)
81b.	Labrum medium or short (width/length ratio >1.9)	82
82a.	Labrum with one tooth (or if with three apparent teeth, front maculation is short and C-shaped, or if front maculation is absent or reduced, rear of elytra with microserrations) (Fig. 4.21B)	83
82b.	Labrum with three teeth (or if only one obvious tooth, front maculation long and angles away from outer edge of elytra) (Fig. 4.21A)	88
83a.	Maculations reduced; first antennal segment with three, sometimes four long, sensory setae	84
83b.	Maculations "normal," confluent, or if reduced, first antennal segment with five long, sensory setae	86
84a.	Rear maculation absent; restricted to southeast Arizona	Cochise Tiger Beetle (*Cicindela pimeriana*)
84b.	Characters not as above; only rear maculation present	85
85a.	Elytral surface appears greasy with distinct pits; body larger; more distinct microserrations along elytral edge; rear maculation smaller with front edge horizontal	Williston's Tiger beetle (*Cicindela willistoni praedicta*)
85b.	Elytral surface appears smooth with small, indistinct pits; finer, less distinct microserrations along elytral edge; rear maculation larger with front edge angled forwards toward inner edges of elytra	Great Basin Tiger Beetle (*Cicindela amargosae*)
86a.	Elytral surface dull (with many tiny pits between punctures)	87
86b.	Elytral surface appears greasy with distinct pits and bumps	Williston's Tiger Beetle (*Cicindela willistoni*)
87a.	Combined tarsal segments of hind leg as long as its tibia; front maculation absent or if present not reaching as far toward middle as bend of middle maculation (Fig. 4.32A)	Senile Tiger Beetle (*Cicindela senilis*)
87b.	Combined tarsal segments of hind leg shorter than its tibia; front maculation reaching as far toward middle as bend of middle maculation (Fig. 4.32B)	Short-legged Tiger Beetle (*Cicindela tenuicincta*)

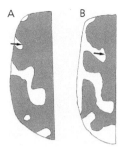

Figure 4.32 A, Front maculation does not reach as far toward the inner edge of the elytra as does the bend of the middle maculation, Senile Tiger Beetle (*Cicindela senilis*); **B**, front maculation reaches as far toward the inner edge of the elytra as does the bend of the middle maculation, Short-legged Tiger Beetle (*C. tenuicincta*).

88a.	Maculations reduced, middle maculation absent or reduced to a triangle; elytra reddish; large body size (>15 mm)	Beautiful Tiger Beetle (*Cicindela pulchra*)
88b.	Maculations "normal" or if reduced, elytra not red; medium body size (<15 mm)	89
89a.	Elytra shiny, with deep pits often reddish	Crimson Saltflat Tiger Beetle (*Cicindela fulgida*)
89b.	Elytra dull, with shallow pits or granules and punctures	90
90a.	Front maculation complete or at least with a dot on shoulder	Oblique-lined Tiger Beetle (*Cicindela tranquebarica*)
90b.	Front maculation absent, no dot on shoulder	91
91a.	Great Basin from Central Oregon to British Columbia; surface of elytra less shiny; labrum of female black	Sagebrush Tiger Beetle (*Cicindela pugetana*)
91b.	Southeastern Oregon and northern California to Utah and western Montana; surface of elytra more shiny; labrum of female black	Alpine Tiger Beetle (*Cicindela plutonica*)
91c.	Restricted to Death Valley area of eastern California; labrum of female whitish	Oblique-lined Tiger Beetle (*Cicindela tranquebarica arida*)
91d.	Restricted to San Joaquin Valley of California	Oblique-lined Tiger Beetle (*Cicindela tranquebarica joaquinenesis*)

Identification of Tiger Beetle Larvae

The larval stages of tiger beetles have received much less interest and study than adults, probably because of their burrow-dwelling habits and aesthetically less pleasing grub-like appearance. Finding the burrows of larvae requires examination of the ground surface at close range (usually you have to hunch over a bit to see them), a very different search image than that of adults where you need to be looking 5 to 20 m ahead. Another problem is the difficulty in identifying larvae to species. The characters used (especially numbers and arrangement of setae on the pronotum and other body parts and characteristics of the hooks on the fifth abdominal segment) are often difficult to find and less distinctive than most adult characters. Also, these characters may change as the larvae advance through each of their three distinct growth stages (instars). This historic disregard for larvae is unfortunate because biologically and ecologically larvae often are more important than adults as habitat indicators and for evaluating the viability and dynamics of populations.

Another problem with studying and identifying larvae is that only about 60% (70 of 117) of the species from Canada and the United States are described, and some of the descriptions are incomplete. Most of these descriptions are of only the third instar, which is the basis of most larval taxonomy. Virtually no one has studied the differences among subspecies or variation within a species. There is no comprehensive key to the described larvae, and most regions and habitats of Canada and the United States, especially in the west, include many species for which the larvae have not been described. A few regional keys to third instar larvae are available for small parts of the United States, but they are not useful for large parts of the continent. Despite these problems, larvae may often be identified with some certainty by using other information, such as knowledge of the adults present in the larval habitat, the larval instar number, diameter of the larval burrow or of the larval pronotum, depth and microhabitat of the burrow, and peculiar construction of the burrow.

Because of this lack of information for many species of larvae, we cannot include descriptions or keys to this level. However, larval characteristics of the genera found in the United States and Canada are often distinct (Fig. 5.1). Using these characters, we provide a comparative description and a simple key to the major taxonomic groups of tiger beetles of the region.

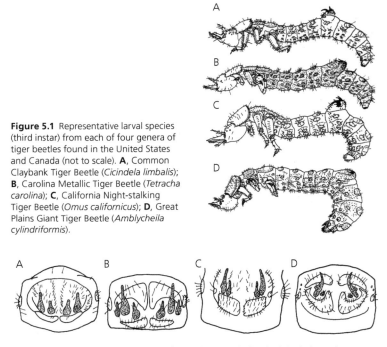

Figure 5.1 Representative larval species (third instar) from each of four genera of tiger beetles found in the United States and Canada (not to scale). **A**, Common Claybank Tiger Beetle (*Cicindela limbalis*); **B**, Carolina Metallic Tiger Beetle (*Tetracha carolina*); **C**, California Night-stalking Tiger Beetle (*Omus californicus*); **D**, Great Plains Giant Tiger Beetle (*Amblycheila cylindriformis*).

Figure 5.2 Fifth abdominal segment of larva showing the hooks (stippled areas) on a representative species of each of the four genera of tiger beetles found in the United States and Canada. **A**, Giant Tiger Beetle, Genus *Amblycheila*, **B**, Night-stalking Tiger Beetle, Genus *Omus*; **C**, Metallic Tiger Beetle, Genus *Tetracha*, **D**, Common Tiger Beetle, Genus *Cicindela*.

Giant Tiger Beetles (*Amblycheila*): Second pair of simple eyes distinctly smaller than first pair; inner and median hooks on the fifth abdominal hump similar in shape, thorn-like and with numerous stout setae; inner and median hooks distinctly separated at their bases (Fig. 5.2A); fourth antennal segment very small, less than one-fifth the length of the third segment. Night-stalking Tiger Beetles (*Omus*): Second pair of simple eyes distinctly smaller than first pair; fifth abdominal hump with three pairs of hooks: inner, median, and lateral hooks; median pair longer than the other two (Fig. 5.2B).

Metallic Tiger Beetles (*Tetracha*): Second pair of simple eyes not less than one-half the diameter of the first pair; inner and median hooks on fifth abdominal hump similar in shape, thorn-like, and with fewer and more slender setae; inner hooks much smaller; inner and median hooks nearly touching at their bases (Fig. 5.2C); fourth antennal segment only slightly shorter than third segment.

Tiger beetle genera in the Tribe Cicindelini: First and second pair of simple eyes slightly different in size; median hooks on fifth abdominal segment long, curved, and sickle-shaped and directed outward; inner hooks short, cylindrical and usually with a short sharp spine (Fig. 5.2D).

Key to the Third Instar Larvae of the Genera of Tiger Beetle Found in Canada and the United States

1a.	Hump on the dorsal side of fifth abdominal segment with three pairs of hooks, middle pair distinctly longer than other two (Fig. 5.2B)	Night-stalking Tiger Beetles (*Omus*)
1b.	Hump on the dorsal side of fifth abdominal segment with two pairs of hooks	2
2a.	The two pairs of hooks on the hump of the fifth abdominal segment different in shape, median pair long, and curved; inner pair short and cylindrical (Fig. 5.2D)	Tiger beetle genera in the tribe Cicindelini
2b.	The two pairs of hooks on the hump of the fifth abdominal segment similar in shape, thorn-like, but inner hooks distinctly shorter	3
3a.	Inner and middle hooks close together and nearly touching at their bases (Fig. 5.2C)	Metallic Beetles (*Tetracha*)
3b.	Inner and middle hooks distinctly separated at their bases, not touching (Fig. 5.2A)	Giant Tiger Beetles (*Amblycheila*)

Species Accounts

We use Freitag's *Catalogue of the Tiger Beetles of Canada and the United States*, Erwin and Pearson's *A Treatise on the Western Hemisphere Caraboidea*, Vol. 2, Duran and Gwiazdowski's *Systematic revision of Nearctic Cicindelini*, and Wiesner's *Checklist of the Tiger Beetles of the World* as a taxonomic basis for the status of names. Any deviations from these classification schemes will be indicated and justified.

We also use the most widely accepted common names of tiger beetle species in conjunction with scientific names. Subspecies, however, are indicated only by their scientific moniker.

Distribution Maps

Using a compilation of published regional records and collection data from specimens in museums and personal collections, we assembled the most complete and current distributional status possible for each tiger beetle species and recognized subspecies occurring within the political boundaries of the United States and Canada. For convenience, we have illustrated these geographic ranges with area maps, and state and province boundary lines help specify each range map. All subspecies for each species are included on the same map and distinguished with different patterns of shading. Areas of subspecies intergradation are indicated with overlapping shading types. Note that we do not refer to the intergrades between subspecies as "hybrids." We reserve this term for cases when two normally separate species produce offspring.

Each of these distribution maps is an approximate outline defined by records at the extreme edges of the species' distribution. Please note, however, that no species will occur everywhere within the indicated range. There are many areas within a distribution where the specific habitat for that species does not exist and from which it is thus absent. However, only when a species is absent from significantly large intermediate areas will we indicate these absences on the range map. In some instances, however, it is more appropriate to represent the distribution or parts of the distribution with individual symbols (circles, triangles or diamonds). Species restricted to narrow coastal distributions and those that have extremely specialized and disjunct habitat requirements are more conveniently presented in this manner. Similarly for isolated colonies and vagrant individuals we use individual symbols with arrows indicating probable routes of dispersion.

Giant Tiger Beetles, Genus *Amblycheila*

The seven species of this strictly nocturnal genus are confined to the western United States and northern Mexico. Five species occur north of Mexico, and two, *A. nyx*, from Quatrocienegas Bolson in west central Coahuila, and *A. halffteri* from San Luis Potosi, are endemic to northeastern Mexico. Adults of this genus are flightless with fused elytra, all dark brown to black and are the largest tiger beetles in the Western Hemisphere. Each species occupies a distinctive habitat, and no two species are known to occur together at the same site. Giant Tiger Beetles are not attracted to lights at night.

Montane Giant Tiger Beetle, *Amblycheila baroni* Rivers
(Plate 1) [Map 1]

Description and similar species: Length: 20–25 mm; all black to dark brown with a smooth but dull or matte surface on the elytra; a single (rarely two), indistinct, raised pleat (carina) runs the length of the elytron just above the outer edge (Fig. 4.4A). Shallow pits (punctae) on the elytral surface are indistinct. Other species of this genus are similar in shape, and color, but they all are larger, have a shiny surface to the elytra and/or three or more distinct pleats (carinae) on each elytron.

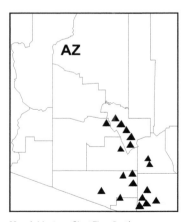

Map 1 Montane Giant Tiger Beetle, *Amblycheila baroni*.

Subspecies and morphological variants: No distinct subspecies or geographical variants are recognized.

Distribution and habitats: Restricted to open pinyon–oak–juniper areas of southern and central Arizona at mid-elevations above 900 m with sand to gravel substrate and huge granite boulder fields. A possible record from similar habitat in west Texas near Big Bend National Park may indicate that this species has a much larger range than now known. Establishing its range accurately, however, is made difficult by a highly disrupted distribution. It is strangely absent from what appear to be otherwise ideal habitats, such as the Pajarito Mountains of south central Arizona and the Chiricahua Mountains of southeastern Arizona. The similar Mojave Giant Tiger Beetle occupies similar habitat to the west, and the ranges of the two species may overlap in low desert mountains of west central Arizona.

Behavior: Adults are completely nocturnal and flightless. They can most frequently be seen running across roads through appropriate habitat and along the base of the rounded and huge boulders that are a typical part of their habitat.

Seasonality: Adult activity is limited to the warm, wet monsoon season from July to September.

Larval biology: The huge larvae occur in the same habitat as adults, and have been found under flat rocks and around the bases of large rocks and boulders.

Mojave Giant Tiger Beetle, *Amblycheila schwarzi* W. Horn
(Plate 1) [Map 2]

Description and similar species: Length 21–27 mm; all black to dark brown with a shiny surface on the elytra, three distinct, raised pleats (carinae) run the length of the elytron, two along the near-vertical side and the inner one delineates the flat horizontal center of the elytron from the sloping side (Fig. 4.4B). Several rows of distinct shallow pits run along the sloping elytral edge and two irregular rows run along the inside of each central pleat. The central flat area of the elytra lack distinct pits. Other species of this genus are similar in shape, and color, but they have only one pleat (Montane Giant Tiger Beetle), or if three pleats, different geographic distributions (Plateau Giant Tiger Beetle in the southern intermontane Rocky Mountain area and South Texas Giant Tiger Beetle) or many rows of pits cover the entire center of the elytra (Great Plains Giant Tiger Beetle).

Subspecies and morphological variants: No distinct subspecies are described for the Mojave Giant Tiger Beetle, however specimens from southern Utah average 2–3 mm shorter than those from other locations within its range.

Distribution and habitats: Restricted to open live oak and juniper areas with large boulder fields and sandy crevices between 1000 and 1500 m elevation in

Map 2 Mojave Giant Tiger Beetle, *Amblycheila schwarzi.*

the Mojave Desert. Although it overlaps geographically with the Plateau Giant Tiger Beetle, this latter species is most often found in lower elevation grassland areas. There is a possibility that the Montane Giant Tiger Beetle and Mojave Giant Tiger Beetle may occur together in low mountains of central western Arizona.

Behavior: Adults are completely nocturnal and flightless. They can most frequently be seen running in dry washes and along the base of the rounded and huge boulders that are a typical part of their habitat, especially after rains.

Seasonality: Adult activity is primarily in the spring April to June and again in August.

Larval biology: Larva unknown.

Great Plains Giant Tiger Beetle, *Amblycheila cylindriformis* Say (Plate 1) [Map 3]

Description and similar species: Length 29–35 mm; all black to dark brown with a shiny surface on the elytra that has many closely spaced rows of distinct pits over the entire surface. Three distinct, raised pleats (carinae) run the length of each elytron, two along the near-vertical sides and the inner one on the upper part of the elytron. Other species of this genus are similar in shape, and color, but no other Giant Tiger Beetle species has the entire surface of the elytra covered with rows of pits.

Subspecies and morphological variants: No distinct subspecies or geographical variants are known.

Map 3 Great Plains Giant Tiger Beetle, *Amblycheila cylindriformis*.

Distribution and habitats: Limited to grassland and prairie areas, pastures and well-drained bare or sparsely vegetated soil. This species is found between 540 and 1460 m in the Great Plains. It has also been observed on clay banks near creeks, rivers, and ravines as well at the tops of low cliffs.

Behavior: Adults are completely nocturnal and flightless. They can most frequently be seen running across roads and through open patches of bare earth in grasslands. Adults pass the daylight hours underground, often in rodent burrows.

Seasonality: Adult activity is from April to September in the southern part of its range and June to August in the northern part of its range.

Larval biology: The huge larvae occur in the same habitat as adults, and they have been found in the fresh dirt at the mouths of prairie dog, badger and gopher tunnels as well as on near vertical faces of clay bluffs. They tend to occur in clusters of 2 to 10 individuals within a 25 cm radius, and they forage from the mouths of their tunnels mainly after sunset but also on cloudy days.

South Texas Giant Tiger Beetle, *Amblycheila hoversoni* Gage (Plate 1) [Map 4]

Description and similar species: Length 32–36 mm; the largest tiger beetle in the Western Hemisphere, the South Texas Giant Tiger Beetle was only described in 1990. It has a shiny black head and thorax that usually contrast with the shiny but reddish to dark brown elytra. Three distinct, raised pleats (carinae) run the length of each elytron just above the outer edge, and two run along the near-vertical outer edge and the inner one on the upper part of the elytron. Shallow pits on the elytral surface are distinct and form two regular rows along the central portion of each elytron. Other species of this genus are similar in shape, and color, but peculiar to the South Texas Giant Tiger Beetle is a single row of pits between the inner two pleats. All other Giant Tiger Beetle species with three pleats have two or more rows of pits here.

Subspecies and morphological variants: No distinct subspecies or geographical variants are known.

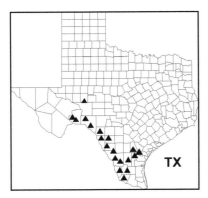

Map 4 South Texas Giant Tiger Beetle, *Amblycheila hoversoni.*

Distribution and habitats: Known only from south and west central Texas, this species occurs most commonly on the floor of undisturbed thorn tree habitat with good drainage and covering low rolling hills with caliche soils. In more open and disturbed habitats of west Texas, it occurs in much lower numbers. It has not yet been collected in Mexico but almost certainly occurs in similar habitat south of the Rio Grande.

Behavior: Adults are completely nocturnal and flightless, and they can most frequently be seen running quickly through leaf litter and near animal burrows after midnight. During the daytime they retreat into burrows.

Seasonality: Adult activity extends from April to November, except June. They are most active following rainy periods in the spring and autumn.

Larval biology: Larvae that are undoubtedly of this species have been found in eroded hillsides and in the upper surfaces, or ceilings, of badger and armadillo burrows.

Plateau Giant Tiger Beetle, *Amblycheila picolominii* Reiche (Plate 1) [Map 5]

Description and similar species: Length 24–28 mm; all black to dark brown with a shiny surface on the fused elytra. Three distinct, raised pleats (carinae) run the length of each elytron just above the lateral margin, two along the near-vertical outer edge and the inner one on the upper part of the elytron. Shallow pits on the elytral surface are distinct, with a single row running along the inner length of each inner pleat. Most similar to the South Texas Giant Tiger Beetle, the distinguishing character for the Plateau Giant Tiger Beetle is a smaller and less robust size and three or four rows of pits between the inner two pleats on the elytra, one of which is made up of large pits and the others of small pits. The South Texas Giant Tiger Beetle is much larger and has but one row of large pits between these pleats.

Subspecies and morphological variants: No distinct subspecies or geographical variants are known.

Map 5 Plateau Giant Tiger Beetle, *Amblycheila picolominii*.

Distribution and habitats: The Plateau Giant Tiger Beetle replaces the Great Plains Giant Tiger Beetle in dry open habitats of the southern intermontane Rocky Mountain area. Here it occurs in grassy areas to open juniper woodland with scattered rocks and small boulders. In southern New Mexico it has been found in salt flats and low saline areas.

Behavior: Adults are completely nocturnal and flightless. They most frequently can be seen running across roads and through open patches of bare earth in grasslands.

Seasonality: Adult activity is limited to the warm, wet monsoon season from June to August in New Mexico and from July to September in Arizona.

Larval biology: An individual of what is most likely this species was observed at the mouth of its tunnel in soft dirt surrounding a gopher hole in the bottom of Canyon de Chelly in northeastern Arizona.

Night-stalking Tiger Beetles, Genus *Omus*

Resembling several species of closely related ground beetles in other tribes of the family Carabidae, *Omus* is the most poorly known genus of tiger beetles in North America. More than 100 species and subspecies of Night-stalking Tiger Beetles have been described, but a paucity of clear distinguishing characters makes it difficult to determine taxonomic differences. However, some distinctions in male genitalia and molecular analysis hold hope for taxonomic clarifications in future studies.

Presently, various authors place them into five to fifteen species with varying numbers of indistinct subspecies. The biology and activity periods for many of these populations have yet to be studied; some are known from only a few specimens, and accurate distribution records for several populations are unavailable. All are black to dark brown, large to medium-sized, and flightless with fused elytra. They are primarily nocturnal but are occasionally active on overcast days. They occupy the Pacific coastal lowlands, slopes of coastal mountains, and the interior west slopes of the Cascade/Sierra Nevada ranges. Two or more species occur together in several localities. The life cycle of most species is apparently 3 or more years.

Greater Night-stalking Tiger Beetle, *Omus dejeani* Reiche
(Plate 2) [Map 6]

Description and similar species: Length 18–21 mm; dull black with bronzy reflections, the Greater Night-stalking Tiger Beetle is the largest species in the genus. Defined ledges extend out from the upper sides of the thorax to give it a bulky appearance and a distinctively triangular shape. The elytra are only moderately domed in profile, and shallow but broad dimples on their surfaces are unique among the *Omus*. All other species in this genus have smoothly granulated elytral surfaces that, in profile, are more steeply domed.

Subspecies and morphological variants: No distinctive differences in size or body characters have been noted for any geographical populations.

Distribution and habitats: The Greater Night-stalking Tiger Beetle occurs on the floor of relatively tall temperate rain forests and adjacent grasslands from the coast to 300 m elevation in coastal mountains and the Cascades of

Map 6 Greater Night-stalking Tiger Beetle, *Omus dejeani.*

northwestern United States and southwestern western British Columbia (including the southern half of Vancouver Island). Occasionally on upper ocean beaches under drift wood. In coastal forests, it regularly occurs together with Audouin's Night-stalking Tiger Beetle.

Behavior: During the day, this species is active under forest floor litter and emerges into the open only on cloudy days. During the night, individuals hunt for prey in open areas and seek bare clay banks at the forest edge on which to oviposit. Commonly captured and eaten by several species of shrews and skunks.

Seasonality: Most active during warmer months from March to October, but can be active on warm days of the winter as well, especially in the southern and lower elevation portions of its range.

Larval biology: Typically clusters of burrows in vertical banks of clay or on trails. The burrows on vertical soils remain open and active all through the year and are 15–20 cm deep.

Lustrous Night-stalking Tiger Beetle, *Omus submetallicus* G. Horn (Plate 2) [Map 7]

Description and similar species: Length 15–18 mm; highly endemic and poorly known, this species is similar to the California Night-stalking Tiger Beetle but under magnification in the hand, the Lustrous Night-stalking Tiger Beetle has a unique row of stout, hair-like setae along each upper side of the thorax (Fig. 4.5).

Subspecies and morphological variants: Except for the highly localized type location no additional populations are known.

Distribution and habitats: Know only from extreme western Fresno County on the eastern slopes of the coastal range (Diablo Range) facing onto the San Joaquin Valley in central California, this species occupies the driest habitat of any Night-stalking Tiger Beetle. It has been found only in open Blue Oak woodlands and Digger Pine forest floor along a 10 km corridor of mostly

CA

Map 7 Lustrous Night-stalking Tiger Beetle, *Omus submetallicus.*

north-facing slopes of canyons and stream cuts bordered on the west by the mouth of Wartham Canyon (600 m elevation). It has most commonly been caught in pit falls, but it also hides under Digger pine cones, dried cow dung and logs. Many workers have mistakenly identified *Omus californicus lecontei* as *O. submetallicus* because it occurs within meters of *O. submetallicus* in the western end of Wartham Canyon.

Behavior: Little is known of this species behavior, but it apparently is nocturnal and most commonly observed on steep slopes where it likely runs under deep leaf litter.

Seasonality: Adults have been caught in pit fall traps from mid-March to early June.

Larval biology: Burrows have been found in same microhabitats as adults in moist sticky clay soil on shaded slopes covered with oak leaf detritus and stream cuts. The larvae plug their burrows and become inactive by early summer when it becomes hot and dry.

California Night-stalking Tiger Beetle, *Omus californicus* W. Horn (Plate 2) [Map 8]

Description and similar species: Length 12–18 mm; more than 80 forms have been described that are tentatively considered the same species as *O. californicus*. Highly variable in size and elevational range, it is generally distinguished from other species of Night-stalking Tiger Beetles by its domed and granulate elytral surface and by a relatively flat thorax, whose forward corners extend to the side but are not distinctly turned down (Fig. 4.6A). Under a microscope, the pitted areas of the elytral surface are regularly spaced and separated by intermediate smooth areas that are equal to or wider in width. The Lustrous Night-stalking Tiger Beetle is similar but has a distinct row of stout, hair-like setae along each upper side of the thorax (Fig. 4.5).

Subspecies and morphological variants: Low elevation coastal forms are generally considered the nominate subspecies. Those from higher elevations in the Sierra Nevada are considered by some as subspecies **intermedius Leng**. The populations at intermediate elevations between the coast and Sierra Nevada in northern California have been given the subspecies name **angustocylindricus W. Horn**. Although distinguished by elevation and a few behavioral differences, no consistent morphological characters readily separate these forms. Considerable taxonomic confusion remains, and this complex of populations is in need of detailed molecular, morphological and ecological studies to sort them out with confidence.

Map 8 California Night-stalking Tiger Beetle, *Omus californicus*; **A**, *O. c. californicus*; **B**, *O. c. angustocylindricus*; **C**, *O. c. intermedius*.

Distribution and habitats: Except for a few sites in extreme southwestern Oregon, the California Night-stalking Tiger Beetle is restricted to California. Here it occupies densely shaded coniferous groves, especially Redwoods, along streams west of the coastal ranges below 900 m south to Southern California. In some of the southern parts of its coastal range, it is found sparingly in oak woodlands. It is absent from the San Joaquin Valley but occupies the west slope of the Sierra Nevada from northern California south to central California up to 2400 m. At high elevations it is found on the floor of coniferous forest, especially Giant Sequoia groves. In Shasta County, California it has been found at intermediate elevations on the floor of Ponderosa and Digger Pine forests, suggesting a transverse connecting corridor between the coastal and montane populations.

Behavior: Coastal and lower elevation populations have typical nocturnal and cloudy day activity patterns running in leaf litter. At higher elevations it occupies more open soil substrates and is generally crepuscular, even at near freezing temperatures for a few hours. On colder nights it quickly retreats to under rocks or burrows and remains inactive.

Seasonality: Adults in coastal habitats are active from March to June in the northern part of its range and from January to May in the southern part. In the Sierra Nevada adults are active from late April to June.

Larval biology: Burrows often clustered in moist clay soil along paths and trails through redwood groves.

Audouin's Night-stalking Tiger Beetle, *Omus audouini* Reiche (Plate 2) [Map 9]

Description and similar species: Length 14–18 mm; dull black, Audouin's Night-stalking Tiger Beetle is variable in size but distinguished from other *Omus* species with domed elytra by the forward corners of the thorax curved distinctly downward (Fig. 4.6B). Under a microscope, the small pits on the surface of the elytra are arranged in an irregular pattern with relatively wide pitted areas divided by narrower smooth areas.

Subspecies and morphological variants: Fifteen forms of *O. audouini* have been described as separate species or as subspecies of *O. audouini* or *O. californicus*, but they are distinguished primarily by geographic location with no consistent morphological or behavioral differences. In general, individuals in the north are larger with a more pronounced texture to the elytral surface, and they gradually become smaller and smoother toward the south.

Map 9 Audouin's Night-stalking Tiger Beetle, *Omus audouini*.

Distribution and habitats: Primarily occupies shaded and moist forest floor leaf litter of the coastal plain, but Audouin's Night-stalking Tiger Beetle also forages out into high grassy areas and clay banks above the ocean. Generally it occurs in more open areas than the Greater Night-stalking Tiger Beetle, but in moderately shaded forests at low elevations, such as in western King and Pierce counties, Washington, the two species can be found together. However, specimens have been found in moist canyons along the Columbia River as far east as Benton County in central south Washington and east of the Cascades in Klamath County, Oregon. As recognized presently, Audouin's Night-stalking Tiger Beetle occurs from southwestern British Columbia south to northwestern California.

Behavior: Primarily active at night, Audouin's Night-stalking Tiger Beetle can be found foraging under leaf litter or in open areas such as clay road cuts and coastal bluffs on cloudy days. It is readily captured in pitfall traps.

Seasonality: Adults are active from April to June in the northern parts of its range and from February to May in the southern part.

Larval biology: Burrows in clay soil of horizontal bare areas. Not active during the winter months.

Mount Ashland Night-stalking Tiger Beetle, *Omus cazieri* *van den* Berghe
(Plate 2) [Map 10]

Description and similar species: Length 14–16 mm; described in 1994, this species is similar to Audouin's Night-stalking Tiger Beetle. The forward corners of the thoracic pronotum are curved distinctly downward (Fig. 4.6B), but the surface microsculpturing on the elytra of the Mount Ashland Night-stalking Tiger Beetle is deeper and more distinct. Under a microscope, it is most readily distinguished from other species of Night-stalking Tiger Beetles by the long and stout tip (apical lobe) of the male aedeagus. No other Night-stalking Tiger Beetle species except the Greater Night-stalking Tiger Beetle has an aedeagus tip as heavy.

Subspecies and morphological variants: No distinct populations of this highly endemic species have been described, but because this population is so poorly studied its relation to other nearby populations of *Omus* needs to be clarified.

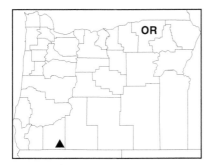

Map 10 Mount Ashland Night-stalking Tiger Beetle, *Omus cazieri.*

Distribution and habitats: Known only from Mt. Ashland and vicinity, Jackson County, in southwestern Oregon, the Mount Ashland Night-stalking Tiger Beetle is found during the day under leaf and pine needle litter on the floor of mixed conifer forest dominated by Douglas Fir. On cloudy days and at night it often ventures into adjacent open areas with little vegetation.

Behavior: Almost all specimens of this species have been collected in pit fall traps, and it appears to be limited to nocturnal activity under dense forest floor litter.

Seasonality: Collected in pit falls from February to June

Larval biology: Larva unknown

Metallic Tiger Beetles, Genus *Tetracha*

Until recently, *Megacephala* was considered a Pantropical genus, and its members in the Western Hemisphere were in the subgenus *Tetracha*. Naviaux's revision of this group in 2007, however, restricts the name *Megacephala* to several Old World species. We follow Naviaux's revision, raise the name *Tetracha* to a full genus and apply the English name Metallic Tiger Beetles.

These beetles are primarily nocturnal, and adults of most species are flightless or fly only weakly. Adults of the 110 New World Metallic Tiger Beetle species tend to be highly colorful with bodies of metallic copper, blue and green predominating. Their legs, antennae and mouth parts are usually yellowish. Some species of this genus can hear sounds of prey calling from underground and have been proposed as natural control of lawn pests such as mole-crickets.

Several species range as far south as northern Argentina and Chile, but the greatest diversity of Metallic Tiger Beetles species is in the southern Amazonian region of Bolivia and Brazil. Four species occur in southern portions of the United States.

Carolina Metallic Tiger Beetle, *Tetracha carolina* L.
(Plate 3) [Map 11]

Description and similar species: Length 12–20 mm; metallic maroon, green and purple upper surfaces; pale legs, mandibles, and antennae. The broad crescent-shaped maculations at the tip of the elytra expand in width abruptly at their front ends, a unique character. The Virginia Metallic Tiger Beetle is much larger (up to 25 mm), dark, oily green and lacks the pale crescents on the end of the elytra. The Upland Metallic Tiger Beetle reaches North America only in the extreme southern tip of Texas, and it is blackish-green with a relatively smooth elytral surface, and broad, pale crescents on the rear end of the elytra that are not obviously expanded at their front tips. The Florida Metallic Tiger Beetle is darker green with narrow crescents on the rear ends of the elytra.

Subspecies and morphological variants: At least four distinct populations of the Carolina Metallic Tiger Beetle have been described from Mexico and the Caribbean. However, only one of these (*T. carolina carolina*) reaches the United States.

Distribution and habitats: The Carolina Metallic Tiger Beetle ranges across the southern tier of states in the United States south to Guatemala and Cuba. This species is active along sandy and muddy edges of rivers, lakes, estuaries, and temporary ponds, but it can move into moist, grassy upland areas.

Map 11 Carolina Metallic Tiger Beetle, *Tetracha carolina*.

Behavior: Chiefly nocturnal, individuals of this species are often gregarious and scurry away quickly to avoid the light of a moving flashlight. However, they are frequently attracted to the base of permanent lights near water. On warm, cloudy days they are occasionally active during the daytime. Adults have functional wings but fly rarely and then apparently only short distances when exposed to extreme danger. During the day, adults normally burrow into soft and moist substrates or gather under water edge detritus.

Seasonality: Adults are active primarily during the summer months throughout most of the range, but in the desert southwest only during the summer monsoons of July to September.

Larval biology:Larvae of this species have been found in a variety of substrates, including bare clay soil away from water, moist beach sand among sparse vegetation, and hard stony or gravelly soil. The larval burrows are often present in large numbers and are straight and 12–30 cm deep.

Florida Metallic Tiger Beetle, *Tetracha floridana* Leng and Mutchler (Plate 3) [Map 12]

Description and similar species: Length 15–16 mm; the Florida Metallic Tiger Beetle is distinguished by its upperparts typically darker metallic green, and the pale crescent on the rear end of each elytron is much narrower than that of the more widespread Carolina Metallic Tiger Beetle. The similarly dark green Virginia Metallic Tiger Beetle lacks the light crescents on the ends of elytra and is considerably larger. The Upland Metallic Tiger Beetle does not occur in Florida.

Subspecies and morphological variants: Formerly considered a subspecies of the Carolina Metallic Tiger Beetle, the Florida Metallic Tiger Beetle has no other distinct populations.

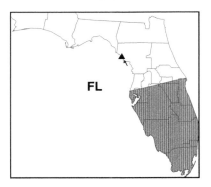

FL

Map 12 Florida Metallic Tiger Beetle, *Tetracha floridana*.

Distribution and habitats: Restricted to the southern half of the Florida peninsula south to Key West, this species occurs primarily in estuarine areas. It occasionally occurs together with the Carolina Tiger Beetle in a narrow band of inland areas in central Florida.

Behavior: Primarily nocturnal, the Florida Metallic Tiger Beetle is occasionally attracted to lights at night. They seldom fly and hide during the day in cracks in the mud, under vegetative detritus and at the base of bushes and grass clumps.

Seasonality: This species is active mainly from May to December.

Larval biology: Larva unknown.

Upland Metallic Tiger Beetle, *Tetracha impressa* Chevrolat (Plate 3) [Map 13]

Description and similar species: Length 15–17 mm; dark metallic green to almost black, the sides of the ivory-colored crescents at the rear end of each elytron are relatively straight with no obvious expansion in width at the front end. Legs, antennae and mandibles orangish-yellow. The other two green species of Metallic Tiger Beetles found in the United States are either much larger with no ivory crescents on the rear end of the elytra (Virginia Metallic Tiger Beetle) or with narrow ivory crescents that are abruptly expanded at their front ends (Florida Metallic Tiger Beetle). The Carolina Metallic Tiger Beetle has a metallic maroon upper body surface.

Subspecies and morphological variants: Formerly considered a subspecies of *Tetracha affinis*, the Upland Metallic Tiger Beetle is now regarded as a separate species with no recognized subspecies.

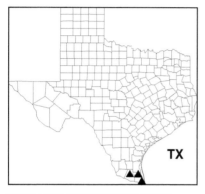

Map 13 Upland Metallic Tiger Beetle, *Tetracha impressa.*

Distribution and habitats: The Upland Metallic Tiger Beetle occurs from southern Texas along the Gulf Coast south to Veracruz, Mexico. Throughout its range the species occupies moist upland areas with water puddles and sparse grass.

Behavior: Seldom flying, the Upland Metallic Tiger Beetle is nocturnal and runs quickly across bare open areas of mud and moist soil. It frequently runs into the base of dense grass clumps to escape danger. Regularly attracted to lights, in Brownsville, Texas, it is frequently found at the base of street lights or around well-lit store windows.

Seasonality: Adults are active mainly during the hot, moist summer from May to September.

Larval biology: Larva unknown.

Virginia Metallic Tiger Beetle, *Tetracha virginica* L.
(Plate 3) [Map 14]

Description and similar species: Length 16–25 mm; dark, oily green, the Virginia Metallic Tiger Beetle is distinguished from all other United States species of Metallic Tiger Beetles by its large size and the lack of ivory-colored crescents on the rear end of the elytra. Legs, antennae and mandibles are dark orangish-yellow.

Subspecies and morphological variants: No subspecies or distinct populations have been described.

Distribution and habitats: Occurs in open grassy areas and along lake and river edges from Wisconsin south to the Gulf Coast. Also common in suburban areas,

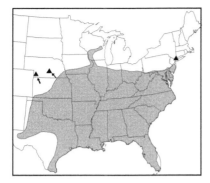

Map 14 Virginia Metallic Tiger Beetle, *Tetracha virginica.*

lawns and open grassy habitats with open bare patches. It apparently enters Mexico only in the extreme northeastern coastal plain of Tamaulipas.

Behavior: Individuals are nocturnal and flightless, and they are regularly attracted to lights at night. During the day they hide under detritus, leaf litter, boards, stones and in mud cracks, often in large numbers together.

Seasonality: Adults are active from late March to November in the southern parts of its range. In the north, it is limited to warm summer months of July and August.

Larval biology: Burrows in lawns, sparse grass in moist clay, sand or gravel. The burrows are relatively shallow (10–20 cm), and with their large diameters (> 6 mm) distinguished from most co-occurring species.

Tribe Cicindelini or Subtribe Cicindelina

This group includes almost 900 of the 2700 known species of tiger beetles worldwide. Adults range tremendously in size (5–25 mm) and color and live in habitats from forest and alpine to desert grassland and ocean beaches. They occur on every continent except Antarctica and all but the most isolated oceanic islands.

A great variety of soil surface types are used as foraging habitat by adults and larvae. Adults of most species in this group forage for food and mate in open areas with little vegetation, but some live on the floor of tropical rain forest. Others are so specialized they occur only on cliff edges, large boulder tops or on fallen logs in moist forest. They quickly respond to predators and danger by flying short distances, although a few are flightless. Primarily on the basis of male genitalia, Rivalier organized the species of this group into 65 subgroups that have been variously considered genera or subgenera.

Here, we follow recent molecular analyses and recognize most of Rivalier's groups as genera. However, these molecular analyses also redefine several of Rivalier's groups. A new genus, *Parvindela,* includes most of the American species that had been previously classified under the largely Old

World genus *Cylindera*. The group *Apterodela* is now elevated to a genus. Previously it was known only from Asia but now includes one American species. Rivalier's *Tribonia* is no longer considered valid. The genus *Eunota* is expanded to include most of what were considered members of the genus *Habroscelimorpha*.

Many species in this group have geographical populations that have been described as subspecies. These subpopulations usually show intermediate forms where their ranges meet. The adult characters that most often are used to distinguish the subspecies include: body color, elytral pattern of maculations, and body size. However, as discussed in Chapter 3, these characteristics may not accurately define evolutionarily distinct groups.

Within some populations individuals with distinctly different forms, such as body color, occur together (polymorphism), and many populations have occasional individuals that exhibit rare color forms (melanism) or unusual elytral patterns. Many species show considerable size difference between the sexes (females generally larger than males), and in a few species, the sexes exhibit noticeable differences in thorax, elytra and mandible shape.

Temperate Tiger Beetles, Genus *Cicindela*

Although there are few external characters that reliably separate this large group of Temperate Tiger Beetles from other groups, Rivalier found male genitalia to be a relatively consistent distinguishing character. The nearly eighty species in it are the most common tiger beetle species at higher latitudes in North America and Eurasia south to northern Mexico and the Middle East, but several species occur as far south as India and South Africa. In North America, except for a few species, almost all members in this subgroup are active as adults in the spring and fall.

Long-lipped Tiger Beetle, *Cicindela longilabris* Say
(Plate 3) [Map 15]

Description and similar species: Length 12–15 mm; the upperparts range in color from dull or shiny black to dark brown, blue and bright green. Both sexes of all forms have a large upper lip (labrum) that is longer than wide. This upper lip shape is unique in North America to the Long-lipped Tiger Beetle. If markings are present on the elytra, they range from three heavy white lines to a single, thin middle line only. The underside of the body varies from dark metallic purple to blue-green or black.

Subspecies and morphological variants: Previously three geographical populations were recognized as subspecies, and one color form, "nebraskana", was considered a separate species. Subsequent molecular and morphological analysis of thousands of specimens has forced us to reconsider these taxonomic divisions. Although some color morphs and elytral patterns are more likely in some geographical areas than others, none of these populations is sufficiently unique to be named as a separate taxonomic group. Considerable variation, polymorphism and intergradation are evident throughout the range of the Long-lipped Tiger Beetle.

The populations found in the boreal zone from the Maritime provinces of Canada and Maine into northwestern Canada are relatively uniform with black above, three maculations on each elytron and metallic blue to green below. However, some individuals on Newfoundland and parts of Quebec and Labrador are green to bronzy above.

In the Rocky Mountains the populations are even more variable, some shiny black, others brown, dark maroon, bronze-green, or olive-green above with wide maculations. The populations in the Cascade, Olympic and Sierra Nevada Mountains vary from black to greenish above and metallic green to coppery below.

Multiple molecular analyses of the "nebraskana" form show that it is a widespread variant. It dominates populations in the Great Basin and western Great Plains. Unlike most other populations in which the long labrum of both sexes is white, the labrum of the female "nebraskana" form is partially or completely dark. The elytra tend to be shinier with at most a thin central maculation on the elytra; the underside is usually black but can be purple or greenish.

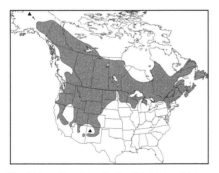

Map 15 Long-lipped Tiger Beetle, *Cicindela longilabris.*

Distribution and habitats: Most commonly found in meadows or grassy areas with open coniferous forests, the Long-lipped Tiger Beetle ranges farther north and at higher elevations in North America than most other tiger beetle species. Its northern limit is probably determined by shallow permafrost that makes larval burrow construction and egg laying in the soil impossible. At higher elevations in the southern Rockies it occurs south to central Arizona, but it is always associated with sandy-gravel, nutrient poor, and acidic soils in montane coniferous forests. In western prairie areas it is found in more nutrient rich soils. In the Sierra Nevada and Cascades ranges, it is often found on heavy volcanic gravelly clays.

Behavior: Active only during the day from mid-morning to late afternoon, the Long-lipped Tiger Beetle is usually not a strong flier, but to escape in areas of closed vegetation, individuals can make vertical flights to heights of 10 m or more. It will often evade capture in flight by suddenly tucking in its wings and quickly falling into dense clumps of tall grasses, where it may be very difficult to locate. But an individual sometimes continues flying down the same path or road track 5 to 10 m at a time. If it flies off into the leaf litter of the forest soil surface alongside the path, it is quickly lost from view because of its camouflage coloring. It is easiest to see on paths, over grown road tracks through open forest, and bare areas within grasslands. In montane areas, it is often found in bare areas created by the tunneling of pocket gophers. This species is not attracted to lights at night.

Seasonality: This species has a spring–fall pattern of adult activity that is altered at higher elevations and latitudes. Adults from eastern boreal zones of Canada and the United States are found from late March to early October, but the greatest activity is in late May and again in mid-August. In the mountains of Washington, Oregon and California, adults are found from mid-May to mid-September but with most activity occurring in early July and again in early September. In the central and southern Rockies adults have been found from mid-May to early September, but most activity is from late July to mid-August. The life cycle is 3–4 years.

Larval biology: Burrows are scattered among open patches often near edges of sparse vegetation on sandy or sandy-clay soils. Burrow depths range from 8 to 20 cm, and the openings often are funnel-shaped.

Bronzed Tiger Beetle, *Cicindela repanda* Dejean
(Plate 4) [Map 16]

Description and similar species: Length 11–13 mm; reddish to bronzy-brown above with three distinct maculations on each elytron that are connected to each other along the outer edge in some individuals and populations. The front maculation is in the shape of the letter "C." Some populations have their maculations reduced to dots and thin lines. The underside is metallic blue-green to coppery and covered with short, white hairs. The constrictions at the front and back of the thorax are similar in width (Fig. 4.29A). Similar species in the east (Twelve-spotted Tiger Beetle) and in the west (Western Tiger Beetle) have less constriction at the front of the thorax than at the back to form a trapezoid shape from above (Fig. 4.29B). These latter two species also have noticeably broader elytra with maculations that are often interrupted to form spots and dots. The front maculation on the Western Tiger Beetle is consistently reduced to two small spots even if the other maculations are complete. Another similar species, the Hairy-necked Tiger Beetle, has the front maculation with a distinctive hook at its rear end forming the letter "G." It also has large tufts of white hairs on the side of the thorax that are absent or less obvious on the Bronzed Tiger Beetle.

Subspecies and morphological variants: Three subspecies of the Bronzed Tiger Beetle are recognized, and their geographical distributions are relatively distinct with narrow zones of intergradation where they meet. Older individuals may be darker, and freshly emerged adults may be temporarily greenish.

C. repanda repanda **Dejean:** The most widely distributed subspecies, the nominate form occurs from the Atlantic Coast almost to the Pacific Ocean in Oregon, Washington and British Columbia. It is bronzy-brown above with thin but complete maculations on the elytra that are usually not connected to each other. However a few individuals in most populations exhibit an incomplete white connection along the outer elytral edge.

C. repanda novascotiae **Vaurie:** Restricted to northwestern Nova Scotia, Prince Edward Island, Cape Breton Island and a few unusual records in southeastern Quebec, this subspecies is more reddish above than other forms of the Bronzed Tiger Beetle. It is distinguished by almost all individuals having their maculations reduced to dots and discontinuous lines.

C. repanda tanneri **Knaus:** Similar to the nominate subspecies, this population is restricted to the Green River Valley of eastern Utah. It is distinguished by extremely broad ivory-colored maculations that are usually connected to each other.

Distribution and habitats: One of the most widely distributed and common tiger beetle species in North America, the Bronzed Tiger Beetle is usually found on sandy beaches along lakes, rivers, and in the east along the ocean. It also occurs in dry sandy areas above beaches as well as in wet roadside ditches, open fields, and in the west on alkaline encrusted soils. It is less common in the west and absent from much of the Great Basin and desert

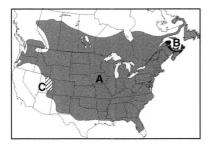

Map 16 Bronzed Tiger Beetle, *Cicindela repanda*; **A**, *C. r. repanda*; **B**, *C. r. novascotiae*; **C**, *C. r. tanneri*.

southwest. Although it extends west along almost the entire stretch of the Columbia River and some of its tributaries, this species only reaches the Pacific coastal plain in central Oregon.

Behavior: The Bronzed Tiger Beetle is diurnal and not attracted to night lights. It is often gregarious and swarms in large numbers along the water's edge of sandy beaches where they pursue small insect prey, and males search for females. Adults dig shallow burrows in sandy soils during the evening and reemerge the following morning. This is one of the few species of tiger beetle recorded eating small fruits in addition to its more common insect fare. The other similar species most likely to be found in the same habitat are the Western Tiger Beetle and the endemic Columbia River Tiger Beetle in the west, and the Twelve-spotted Tiger Beetle, the Hairy-necked Tiger Beetle, and the endemic Appalachian Tiger Beetle in the east and central part of the continent.

Seasonality: In the southeastern United States, adults are active from January to June and again from August to October; in the northeastern part of its range from March to June and then in August to late September; and in the northwestern part of its range from mid-March to early July and mid-August to early October. Some adults, however, are active throughout the summer. Both adults and larvae overwinter. The length of the life cycle is 2 years in most areas.

Larval biology: Burrows, often in high densities, are found at the water's edge in the same open sandy areas as the adults. Larvae often leave the burrow and dig a new one to avoid flooding or desiccating conditions. The larvae are commonly attacked by the parasitoid fly *Anthrax* (Fig. 7.2).

Twelve-spotted Tiger Beetle, *Cicindela duodecimguttata* Dejean (Plate 4) [Map 17]

Description and similar species: Length 12–15 mm; dark brown to blackish above and metallic blue-green below. The maculations range from thin but complete to broken into dots and short lines. The number of these dots often totals 12 and thus the species name. The middle band on the elytra is often thinly connected at a right angle to a short line running along the edge of the elytron. The outer edges of the elytra are bowed out and wide. The front constriction of the thorax is not as deep as the back constriction resulting in a trapezoid shape from above (Fig. 4.29B). It is most easily confused with the Bronzed Tiger Beetle, especially the form *C. r. novascotiae* in the Maritime Provinces of Canada. However, the Bronzed Tiger Beetle has more complete maculations and the thorax is spindle-shaped with equal depth constrictions at the front and back of the thorax (Fig. 4.29A). The middle band of the similar Western Tiger Beetle usually has no connecting line running along the edge of the elytron. Hybrids between the Western and Twelve-spotted Tiger Beetles are common where their ranges overlap on the eastern slopes of the RockyMountains.

Subspecies and morphological variants: Some individuals have no light spots on the elytra. Older individuals can be dark green or even bluish. In the Great Plains, some individuals have complete markings and resemble the Bronzed Tiger Beetle. Along the eastern slopes of the Rocky Mountains, especially in Colorado, this species hybridizes with the Western Tiger Beetle, and a confusing array of individuals bearing intermediate characters is present here. There are no distinct subspecies.

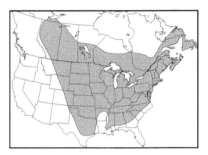

Map 17 Twelve-spotted Tiger Beetle, *Cicindela duodecimguttata.*

Distribution and habitats: Found from the Atlantic Coast west to the Rocky Mountains, this species is absent from much of the southeast and desert southwest. It occurs on moist silty or sandy soils along rivers and lake edges.

Behavior: Often gregarious along water's edge, the Twelve-spotted Tiger Beetle is usually more solitary away from the water's edge. It also commonly scavenges on dead insects.

Seasonality: Adults are active in a spring–fall pattern, usually April to June and August to October. Both adults and larvae overwinter, and the life cycle is 2 years.

Larval biology: Burrows are found in open or sparsely vegetated silty and sandy soils along water's edge with adults. Burrows are often highly aggregated and are 6–10 cm deep.

Western Tiger Beetle, *Cicindela oregona* Dejean
(Plate 4) [Map 18]

Description and similar species: Length 11–13 mm; highly variable in color and size across its range in western North America. The average size of individuals decreases from north to south, but the width of the white maculations increases from north to south. The color above varies from brown, green, purple, and blue to almost black. Most individuals have the same color elytra, thorax and head, but others have the head and thorax of a color different from that of the elytra. Below the color ranges from predominantly metallic coppery in the north to metallic purple in the south. In intermediate areas, many individuals are metallic green below. Where they overlap along the eastern slopes of the Rocky Mountains, especially in Colorado, the Twelve-spotted Tiger Beetle is extremely similar to the Western Tiger Beetle, and hybrids between the two species are confusing. In general, however, the Twelve-spotted Tiger Beetle has a distinct or remnant short white line running along the middle of the outer edge of each elytron and at right angles to the middle band. The Western Tiger Beetle lacks this marginal line. Throughout many of the mountains of the Pacific Northwest and along the coast from Alaska to central California the similar Dispirited Tiger Beetle overlaps with or replaces the Western Tiger Beetle. These two species are quickly separated on the basis of the shape of the middle band on the elytra. In the Western Tiger Beetle this band has a sharp bend toward the rear that forms what is often referred to as an "elbow." In the Dispirited Tiger Beetle, the midline usually has a shallow, wave-like bend with no sharp turn toward the rear (Fig. 4.22).

Subspecies and morphological variants: There are four geographical races presently recognized as subspecies of the Western Tiger Beetle. The subspecies vary in size, body color, and extent of white maculations. The Great Basin, southwestern deserts and the Rocky Mountains apparently have served as partial barriers to gene migration, and the areas of intergradation between subspecies are most evident along these barriers.

C. oregona oregona **Dejean:** The nominate subspecies occurs from the Pacific Ocean inland to the Rocky Mountains. The undersides are metallic blue or purple and the upperside of the head, thorax and elytra are the same color—usually dark brown but with some individuals showing green, blue or rarely purple. The maculations are thin. Intergrades with the subspecies *guttifera* occur in a narrow zone from southern British Columbia south along the crest of the Rocky Mountains to central Utah.

C. oregona guttifera **LeConte:** This subspecies is relatively uniform in color with metallic coppery undersides and uniform rich-brown above. The maculations are moderately heavy. Occurs from central New Mexico to Yukon and interior northeastern Alaska. A narrow zone of intergradation occurs with the nominate form along the crest of the Rocky Mountains, with subspecies *navajoensis* in western Colorado and New Mexico, and with subspecies *maricopa* in western Utah.

C. oregona navajoensis **Van Dyke:** Restricted to a small area along the eastern slopes of the southern Rocky Mountains in Utah, Colorado, Arizona and New Mexico, this subspecies is most similar to the subspecies *guttifera*. The subspecies *navajoensis*, however, is lighter brown above and the maculations are consistently heavier. This subspecies has a narrow zone of intergradation with *guttifera* along the crest of the Rocky Mountains in western Colorado and New Mexico, and with subspecies *maricopa* in southwestern New Mexico.

C. oregona maricopa Leng: The most distinctive of all the subspecies of the Western Tiger Beetle, *maricopa* is metallic dark green to purple below. Above it is usually bicolored with the elytra purple to dark brown and the head and thorax green. The maculations are very broad. Distinct individuals of this form are confined to southeastern and central Arizona. Intergrades with *guttifera* are common in southwestern Utah and with *navajoensis* in southwestern New Mexico.

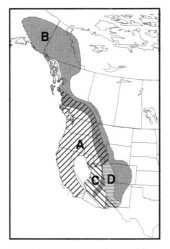

Map 18 Western Tiger Beetle, *Cicindela oregona*; **A**, *C. o. oregona*; **B**, *C. o. guttifera*; **C**, *C. o maricopa*; **D**, *C. o. navajoensis*.

Distribution and habitats: Together with a wide latitudinal and elevational range (sea level to over 3000 m) this species also occupies a wide range of habitats. Along the Pacific Coast it is frequently encountered on or near the ocean beach. Inland in cooler moist areas it is most common along water's edge, such as lakes, rivers, creeks, and reservoirs, but it also ranges commonly into moist upland areas and even sand dunes. In the drier desert regions, populations are entirely restricted to the edges of running streams or reservoirs with sand or mud. Throughout its extensive range, it occurs at the water's edge with numerous other species, especially the Oblique-lined Tiger Beetle, Hairy-necked Tiger Beetle, and Bronzed Tiger Beetle. Along the sandy ocean beaches of Oregon and Washington, however, it occurs with or is largely replaced by the Pacific Coast Tiger Beetle. Along the ocean beaches of central and northern California it occurs with or is replaced by the *eureka* subspecies of the Dispirited Tiger Beetle. Above 2000 m in the Sierra Nevada, Cascades and in the northern Rocky Mountains it is replaced by the nominate form of the Dispirited Tiger Beetle.

Behavior: Often gregarious along confined inland water edges, it is more solitary in upland areas. In the desert southwest, flash floods often can scour the only habitats available for the Western Tiger Beetle in an instant. Populations in this habitat are able to quickly disperse and find appropriate habitat wherever sand and mud have again been deposited. Because of this phenomenon, the *maricopa* subspecies of the Western Tiger Beetle was at one time considered by the US Fish and Wildlife Service for listing as an endangered population. It had disappeared from many of its formerly known breeding sites along Arizona rivers. Further investigation, however, found that flash flood scouring and movement of sand and other substrates force this subspecies to abandon some areas and colonize new ones frequently.

Seasonality: Although adults have been observed active from mid-February to early November, the greatest numbers occur in May–June and again in August–September. In the northern parts of its range, adults show most activity from April to June and August to early October. In Arizona some individuals become active in mid-February, but the greatest numbers are found in April to May and again in September. Adults and larvae overwinter and have a 2-year life cycle.

Larval biology: Burrows are usually clustered in open, moist and loose sand near water on flat ground, but in other areas they are in vertical banks of intermittent streams. Burrows are 12–30 cm deep.

Dispirited Tiger Beetle, *Cicindela depressula* Casey (Plate 4) [Map 19]

Description and similar species: Length 12–16 mm; individuals from coastal areas tend to be larger than those from higher elevations. The species is highly variable in color, especially where it occurs at higher elevations. Above it can be brown, green or blue. Below it can be metallic coppery-green to blue. The maculations are usually complete but thin. Generally the middle band on the elytra has a gradual bend and lacks the sharp "elbow" found in the Western Tiger Beetle. At one time considered a subspecies of the Western Tiger Beetle, later studies showed that the male genitalia of the Dispirited Tiger Beetle are distinct, and that it is a closely related but separate species.

Subspecies and morphological variants: There are two subspecies recognized. One is a Pacific coastal lowland form and the other a high elevation montane form. Considerable intergradation occurs in the Pacific Northwest where the two forms meet at intermediate elevations. An isolated population at Pt. Reyes, Marin County, California, is light brown, larger and present only on sand dunes.

    C. depressula depressula **Casey:** This montane form occurs down to 600 m but is most frequent above 1200 m in the Cascades, Sierra Nevada Mountains, and Rocky Mountains of Montana and Idaho. Individuals show tremendous polymorphism in upperpart coloring with green and blue often the most common, but in some areas brown or even black individuals are regular. The intermediate populations in western British Columbia and adjacent southern Alaska occur at lower elevations down to sea level and tend to be brown. A zone of intergradation also occurs at mid-elevations with *C. d. eureka* in western Washington.

C. depressula eureka Fall: This coastal form is larger and most individuals are almost entirely brown above with older individuals darkening to greenish. The maculations on the elytra are more pronounced than those of the nominate subspecies. The *eureka* subspecies is almost entirely limited to sandy and gravel beaches along coastal plain streams and on the ocean at river mouths opening into the ocean. A few specimens have been found between the coastal ranges and the Cascades of the Willamette Valley in western Oregon.

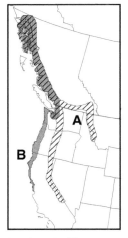

Distribution and habitats: The two forms of this western species occupy a broad range of habitats from coastal sandy beaches to alpine meadows. The lowland forms are restricted to moist sand and gravel in water edge habitats, and the high elevation forms are found in disturbed grassy areas formed by melting snow, recent fires, logging and road cut as well as in alpine areas. At intermediate elevations in western Washington and British Columbia, they are primarily found along gravelly mountain streams.

Behavior: This species can be found alone or in pairs or in groups congregating around moist habitats. It is a strong flier and easy to lose once it flies up from danger.

Map 19 Dispirited Tiger Beetle, *Cicindela depressula*; **A,** *C. d. depressula*; **B,** *C. d. eureka.*

Seasonality: In higher elevations, often just below the snow melt line, the activity of adults is severely limited by the many cloudy, rainy days during the spring season, and thus, although it is a spring–fall species, the spring activity may be masked by inclement weather patterns. Although active from early May to early October, the greater chances for clear warm days in the late summer make it easier to find at this time. The lowland subspecies is active from April to July and again from August to October.

Larval biology: Larvae of the high elevation subspecies have been found clustered in large numbers in a melted snowpack swale at 3000 m in the Sierra Nevada Mountains of California. They were in bare soil patches among the low alpine vegetation. Larvae of the lowland subspecies are yet unknown.

Hairy-necked Tiger Beetle, *Cicindela hirticollis* Say
(Plate 5) [Map 20]

Description and similar species: Length 10–15 mm; across its vast range in North America, this species varies in several characters. Although the color of its upperside is generally brown to reddish-brown, individuals and populations can also be green, black, or blue. Even though the pattern of the maculations is relatively consistent, the heaviness of these maculations differs among populations and often there is considerable variation among individuals within a single population. Two diagnostic characters that almost all individuals of the Hairy-necked Tiger Beetle share are a large tuft of long, white hairs on the side of the thorax and the front maculation shaped roughly in the form of a "G" with a forward hook on its bottom end. It occurs together with many other tiger beetle species on sandy ocean beaches, river, and lake edges, but it is only likely to be confused with the Bronzed Tiger Beetle, which lacks the large tuft of white hairs on the side of the thorax and has its front maculation in the shape of a "C." Also, the mandibles of the Hairy-necked Tiger Beetle are noticeably longer and thinner than those of the Bronzed Tiger Beetle.

Subspecies and morphological variants: A continent-wide study of Hairy-necked Tiger Beetle mitochondrial DNA did not strongly support the validity of any of the named subspecies; however, western populations of the species did exhibit some genetic differentiation. Although the status of most or all of these subspecies is questionable, we include the names of previously recognized subspecies here pending further studies.

Eleven subspecies of the Hairy-necked Tiger Beetle have been recognized, one of which (**C. h. ponderosa** Thomson) is known only from old records from Vera Cruz, Mexico. The subspecies are separated mainly on the basis of mean size, upperside color and elytral pattern, and overall shape of elytra. The three eastern subspecies occupy large areas without major barriers separating them. As a result they are the least distinctive of all populations and show broad zones of intergradation. This great variation within populations makes a determination to subspecies of many individuals unreliable.

In the west, however, populations are more isolated from each other by mountain ranges, deserts and river systems. These subspecies tend to be more distinct and separated on the basis of a few characters. Greenish individuals can occur in almost any population, and they often are very young or very old individuals.

C. hirticollis hirticollis Say: Found east of the Mississippi River and south of the Great Lakes, the nominate subspecies is usually brown to reddish-brown with relatively heavy and almost always complete white maculations. Populations on the Maryland and Virginia barrier islands are significantly smaller than elsewhere. Along the Atlantic seashore on Long Island, New York, and the southern boundary of the Great Lakes it intergrades with the northeastern subspecies *C. h. rhodensis*, which typically is larger, darker brown above and with thin and disrupted maculations. Along the Mississippi River it intergrades with *C. h. shelfordi*, which is the largest of the subspecies and is reddish-brown above with heavy maculations.

C. hirticollis abrupta Casey: Found only in the Sacramento Valley of California, this subspecies is likely extirpated. Above it is dark blackish-brown and the maculations are disrupted. Typically the front maculation is disconnected from the line along the outer elytral edge and its hook-shaped end curls inward more extremely than in other subspecies. Females tend to have broadly widened elytra that easily distinguish them from males. Among subspecies of the Hairy-necked Tiger Beetle, this extreme sexual dimorphism is also evident in *C. h. gravida*.

C. hirticollis athabascensis Graves: This isolated form, found only on the sand dunes bordering Lake Athabasca in northern Alberta and Saskatchewan, is the most northerly of the species. The upper surface of this subspecies ranges from deep blue, green and brown to reddish-brown and purplish. The maculations on the elytra are highly variable but are usually reduced or almost absent. Individuals tend to be large. Because of the actively blowing dry sand in this habitat, the hair-like setae and the outer layers of the cuticle of the bodies of most individuals have been worn down.

C. hirticollis coloradula Graves: Extremely restricted in its range, this subspecies occurs only along the Little Colorado River of northeastern Arizona, just before it enters the east end of the Grand Canyon. Individuals are large and reddish above with very wide maculations. The front maculation usually lacks the tail on the inside and makes it more "C"-shaped. The range of this subspecies is completely surrounded by that of *corpuscula*.

C. hirticollis corpuscula Rumpp: Individuals of this subspecies are similar to *coloradula* but are smaller. Above it is generally reddish-brown with green reflections. The maculations are wide, usually connected to one another, and the front maculation at its lower part ends in a bulb rather than a hook. This form occurs along the Colorado River and its tributaries from southeastern California and southwestern and central Arizona north to southeastern Utah. However, it is now probably extirpated from many former sites along the Gila River in central and western Arizona.

C. hirticollis couleensis Graves: Confined to sandy river banks of the Columbia and Snake Rivers in eastern Washington and Oregon and adjacent Idaho, individuals of this subspecies are intermediate in size. Above they are dark brown to purple with relatively heavy maculations. Reaching west along the Columbia River, *couleensis* extends to Cowlitz County, Washington, within 100 km of the coast, where individuals exhibit some intergrade characters with the coastal form. However, there are no specimens known from the 100 km gap between this western location and the nearest coastal population of the subspecies *siuslawensis*.

C. hirticollis gravida LeConte: Limited to the sandy beaches of the Pacific Ocean from Ensenada, Mexico north to Santa Cruz County, California, this subspecies is largely cut off from coastal forms of the Hairy-necked Tiger Beetle to the north by sea cliffs and rocky areas in northern California unsuitable for this species. The coastal ranges and deserts areas of southeastern California are probably barriers that have isolated *gravida* from the very different *corpuscula* of the Colorado River. Size of *gravida* varies considerably, and the color above is often muddy green, but brownish and bluish individuals occur. The maculations are moderately thick. Females of this subspecies have the sexually dimorphic character of distinctly broadened and curved central edges of the elytra, a character shared among the Hairy-necked Tiger Beetle subspecies only by *abrupta* of the Sacramento River Valley, and which may indicate a genetic relationship. The large population at Pt. Reyes north of San Francisco may be sufficiently isolated to show genetic differences.

C. hirticollis rhodensis Calder: Found throughout the Maritime provinces, New England and the Great Lakes region, this subspecies intergrades considerably with the nominate form. It is generally distinguished, however, on the basis of its large size, darker brown above, and incomplete or thin maculations on the elytra. Some individuals have virtually no maculations.

C. hirticollis shelfordi Graves: This subspecies occupies a large area of the Great Plains from Texas north to the southern portions of the prairie provinces of Canada. Most individuals are large with reddish-brown above. The maculations are heavy and connected to each other along the elytral edge. A narrow zone of intergradation occurs with *corpuscula* along the Green River of eastern Utah, but a broad band of intergradation exists along the Mississippi River with the nominate form.

C. hirticollis siuslawensis Graves: Occurring only on the Pacific ocean beaches at the mouths of rivers from the central Washington coast south to Eureka, California, it is now extirpated from some historic sites. Above it is brown with thin maculations, and the rear hook on the front maculation is absent or quite small.

Distribution and habitats: Widely distributed across North America from coast to coast, this species is rarely encountered away from the immediate

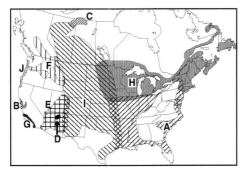

Map 20 Hairy-necked Tiger Beetle, *Cicindela hirticollis*;
A, *C. h. hirticollis*; **B**, *C. h. abrupta*; **C**, *C. h. athabascensis*;
D, *C. h coloradula*; **E**, *C. h. corpuscula*; **F**, *C. h. couleensis*;
G, *C. h. gravida*; **H**, *C. h. rhodensis*; **I**, *C. h. shelfordi*; **J**,
C. h. siuslawensis.

sandy edge of fairly extensive water ways—rivers, lakes, and ocean. However, the most northern subspecies, *athabascensis*, in northern Alberta and Saskatchewan regularly occurs in sand dunes far from water. In Kansas, Oklahoma and Texas, some populations of the subspecies *shelfordi*, occur regularly on moist saline flats.

Behavior: Probably because of its narrow range of moisture tolerance, this species is especially susceptible to droughts, pollution, pesticides, river damming, channelization, shoreline development and destruction of larval habitat by vehicles and other human-caused modifications of its habitat. The species has disappeared from many of its former haunts in New England and the Midwest. No specimens of *C. hirticollis* have been seen or collected in New Hampshire since 1958, even in areas where it used to be common. The coastal Pacific forms can be rare and difficult to locate, especially in southern California, Oregon and Washington. Even many Great Lakes populations have disappeared in the last 50 years. The central Mexican form, *ponderosa*, has not been observed or collected in almost 100 years and may well be extinct. Occasionally individuals are found at lights at night.

Seasonality: Although a spring–fall species, some individuals can be found active from April to October throughout its range. In the north, numbers of adults peak from April to late June and early August to September. In the south it is most common from March to June and then from August to October. Both adults and larvae overwinter. In the southern states, most eggs laid in the spring emerge as adults by August or September, one of the fastest development times known for any species of tiger beetle.

Larval biology: Burrows are restricted to sandy soils near surface water or where the subsurface soil is constantly moist. The density of these typically shallow burrows (8–20 cm) is often high along flood plains, overwash areas and other low-lying water edge sites. The larvae of this species are unusual in that they are regularly found crawling across the soil surface to relocate their burrows in response to changing soil moisture levels.

Sandy Tiger Beetle, *Cicindela limbata* Say
(Plate 6) [Map 21]

Description and similar species: Length 10–12 mm; all but the most northern Canadian populations are characterized by a coalescence of the maculations so that, except for a few small dark patches of the surface, they are mainly white. In Boreal Canada, the maculations are heavy but separated from each other. The dark upperside coloring of the head, thorax and elytral patches is reddish-brown in the north and greenish in the south. The underside varies from coppery-green to metallic purple. The underside and thorax are thickly covered with white hair-like setae. On older individuals these have often been worn away. In its sandy upland habitat few other species occur that can be confused with the Sandy Tiger Beetle. The large Big Sand Tiger Beetle, especially individuals of the *gibsoni* subspecies, also have extensive white areas on the elytra, but they are 3—7 mm longer. In the northern Canadian part of its range, the Sandy Tiger Beetle has a front maculation that is a straight line angled away from the outer edge of the elytron, and the middle maculation bends gradually toward the rear. Both the Blowout Tiger Beetle and Oblique-lined Tiger Beetle share the straight, angled front maculation on the elytra, but both these species have a sharp bend on the middle line.

Subspecies and morphological variants: Five subspecies are now recognized. An additional population, the Coral Pink Sand Dune Tiger Beetle, was considered a subspecies of Sandy Tiger Beetle until it was elevated to a full species based on molecular DNA evidence. Individuals that are greenish above can be present in any population. Most unusual of all, however, are the isolated populations in northwestern Alaska and Quebec, separated by thousands of kilometers from the nearest known populations in central North America.

C. limbata limbata Say: The nominate race occurs in dry sandy areas of eastern Nebraska to adjacent areas of Wyoming and Colorado. The pale area of the elytra is extensive, and the dark upperparts are greenish in most individuals, but dark reddish or bluish individuals are not uncommon. Below it is metallic purplish to dark green. Intergrades with *nympha* to the north are found in intervening areas of South Dakota.

C. limbata hyperborea LeConte: The smallest of the subspecies, this one is also the darkest. Above it is dark reddish-brown with three heavy but separate white maculations on each elytron. The front maculation is a straight and angled line. It is restricted to sandy clearings in pine and poplar forests in the far north of Alberta, Saskatchewan and adjacent Northwest Territories. Intergrade individuals with *nympha* are regular along a narrow band in central Alberta.

C. limbata labradorensis W. N. Johnson: A small and isolated population of *C. limbata* was recently found near the airport at Goose Bay, Labrador. In

1990 it was described as a separate subspecies. Its maculation pattern is most similar to that of *C. l. hyperborea*, with heavy but distinct maculations that are not coalesced. The front maculation is an angled straight line. Almost 3000 km of boreal forest separates this small population from the rest of *limbata*. One hypothesis explains this subspecies as a remnant population isolated as a formerly continuous distribution was disrupted in eastern Canada by climate change and vegetation changes. Alternatively, some individuals were brought here unintentionally on planes flying from central Canada to Goose Bay. The cleared sandy areas around the air strips of this large airport were then ideal for these hitchhikers to establish themselves. Recent observations indicate this form is common in open sandy areas up to 70 km away from the airport. If it can be found in other isolated sandy areas within the boreal forest farther to the west, the "historical isolation" hypothesis will be supported.

C. limbata nogahabarensis Knisley. A large population of this newly described subspecies is known only from the isolated Nogahabara Sand Dunes in north central Alaska, north of the town of Galena and the Yukon River and just south of the Arctic Circle. The next nearest population of Sandy Tiger Beetles, *C. l. hyperborea*, is over 2600 km to the southeast in the Northwest Territories of Canada. Because classical morphology such as elytral color and pattern in this Alaska population varies from dark forms similar to that seen in *C. n. labradorensis* to whitish forms similar to *C. n. nympha*, mitochondrial DNA analysis had to be employed to reveal the phylogenetic and taxonomic relations of this new subspecies.

C. limbata nympha Casey: Found in the extensive sandy areas of North Dakota and Montana north into the prairie provinces of Canada, this subspecies has whitish elytra with a dark reddish-brown wedge extending down the length of their inner borders. A thin dark patch on the back part and occasionally a tiny dark patch on the front part are usually evident on the elytra. The head and thorax are reddish-brown.

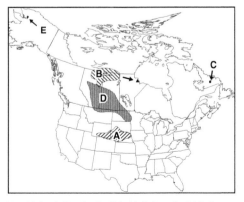

Map 21 Sandy Tiger Beetle, *Cicindela limbata*; **A**, *C. l. limbata*; **B**, *C. l. hyperborea*; **C**, *C. l. labradorensis*; **D**, *C. l. nympha*; **E**, *C. l. nogahabarensis*.

Distribution and habitats: Restricted to dry sandy blowouts, dunes and open sand patches away from water, this species is one of 6 or 7 North American tiger beetle species typical of this habitat. Isolation or expansion of dry sandy areas by climate change, drought, and human intervention of road building, and dryland farming have all probably contributed to the present distribution of this species and its various geographical forms. The dry dust bowl period of the 1930s and 1940s may well have brought the populations of the nominate and *nympha* close enough to have impact on genetic mixing before the subsequent moister period again separated the subspecies.

Behavior: Adults are adapted by both physical characters and behavior to endure extremely high temperatures on the white sand they inhabit. The whitish elytra and white hairs on the underside of the body reflect considerable heat. Adults commonly hide in shady areas or dig burrows to spend the night or escape the highest midday temperatures. This species is not attracted to lights at night.

Seasonality: Adults are active from April to June and then again from early August to late September. However, in many areas the population numbers in the fall are much lower than those of the spring. In the far northern parts of its range, adults are active from June to August. The life cycle in most areas is 2 to 3 years.

Larval biology: Burrows are most common in stabilized sand patches of open or sparsely vegetated interdunal swales or low slopes. They are 20–40 cm deep, and on hot dry days, the larvae plug their burrows to avoid desiccation.

Coral Pink Sand Dune Tiger Beetle, *Cicindela albissima* Rumpp (Plate 6) [Map 22]

Description and similar species: Length 11–12 mm; the elytra are almost entirely ivory-white except for a thin wedge of dark red to reddish-green color that runs down the length of the center. Above the head and thorax are usually greenish-red with a few individuals dark-reddish. It is metallic green below. Easily confused with southern races of the Sandy Tiger Beetle, the Coral Pink Sand Dune Tiger Beetle was until 2000 considered a subspecies of the Sandy Tiger Beetle. Studies of its DNA showed that the Coral Pink Sand Dune Tiger Beetle is a distinct species. Perhaps because of the similar sand dune habitats these two species occupy and the adaptations needed to survive these harsh conditions, superficial similarities in physical structures have evolved in both species to converge on a similar appearance by coincidence. The only other tiger beetle species to occur regularly with the Coral Pink Sand Dune Tiger Beetle is the Oblique-lined Tiger Beetle, which is distinguished by being dark above with maculations reduced to heavy but distinct lines.

Subspecies and morphological variants: No other forms are recognized. Some individuals, however, show an extra small and thin dark patch in the rear part of the otherwise whitish elytra.

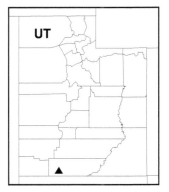

Map 22 Coral Pink Sand Dune Tiger Beetle, *Cicindela albissima.*

Distribution and habitats: Known only from a small area of less than 400 hectares in south central Utah, the entire range of this species is within the borders of the Coral Pink Sand Dunes State Park and the adjacent dune area managed by the federal Bureau of Land Management. Here it occupies dry sand dunes, usually in areas where there is sparse grass.

Behavior: Most individuals occur in lightly vegetated swales of firmer sand in between the dunes and adjacent slopes. They actively forage for small arthropods on the dune slopes but also commonly scavenge dead insects. Adults dig burrows in which they pass the night, inclement days, and high temperatures during early afternoons. Visual censuses of this habitat show the population size to be between 800 and 3000 individuals.

Seasonality: This species exhibits a modified spring–fall activity period, in which most adults are active from late March to May, and only a small number are active from late August to early October.

Larval biology: Burrows are patchily distributed throughout the dune slopes and ridges but most frequent in the sparsely vegetated interdunal swales, in open areas, and among dune grasses and herbs. Burrow depths range from 10 to 40 cm, and larvae are active primarily at night. The burrow openings are plugged with sand in the morning and reopened in late afternoon.

Colorado Dune Tiger Beetle, *Cicindela theatina* Rotger
(Plate 6) [Map 23]

Description and similar species: Length 12–14 mm; individuals are relatively uniform in physical characters. Above it is greenish-brown with coppery reflections. The maculations are very heavy and joined together by a broad white band along the outer edge of each elytron.

Subspecies and morphological variants: No subspecies are recognized, and the most obvious variation within the population is the extent of the white maculations. In some individuals the front maculation is so broad it connects with the curved part of the middle maculation.

Distribution and habitats: Found only on or near the sparsely vegetated borders of the Great Sand Dunes National Park in Alamosa, Costilla and Saguache

Map 23 Colorado Dune Tiger Beetle, *Cicindela theatina*.

counties of south central Colorado, the Colorado Dune Tiger Beetle is restricted to the base of large sand dune fields.

Behavior: Adapted for hot and abrading sand, this species seeks refuge in adult burrows during windy, cold and abnormally hot days. The adult burrows are also used at night.

Seasonality: Although technically a spring–fall species, adults are rarely found before early June, and some individuals can be found throughout the summer. It again becomes more common in August to late September. Both larvae and adults overwinter.

Larval biology: Burrows are found along the sparsely vegetated edges of dune blowouts.

St. Anthony Dune Tiger Beetle, *Cicindela arenicola* Rumpp [Plate 6) [Map 24]

Description and similar species: Length 11–13 mm; above it is coppery-red to greenish-red. The maculations are so broad and coalesced that the individual maculations are obscured. Dark coloring is reduced to a large wedge down the length of the middle of the elytra and a small thin patch between the last and middle maculations. The isolated population in southwestern Idaho (Bruneau Sand Dunes in Owyhee County) was considered a geographic variant of the St. Anthony Dune Tiger Beetle until recently. Detailed studies of genitalia and anatomy show this western population to be a separate species, the Bruneau Dune Tiger Beetle, and it is distinguished by upper parts greenish in color, consistently more expanded and confluent white maculations, and a peculiar tooth on the end of mandible which is projected upward out of alignment with the rest of the teeth (Fig. 4.28).

Subspecies and morphological variants: The upperparts of individuals on small dunes in the western part (Minidoka County) of the small range of the St. Anthony Dune Tiger Beetle tend toward reddish-green. They also show other characters that, except for the lack of the peculiar displaced tooth on the mandible, are intermediate between the St. Anthony Dune Tiger Beetle and the Bruneau Dune Tiger Beetle.

Distribution and habitats: Mainly restricted to sparsely vegetated swales between active sand dunes of the Snake River Plain, which drains into the Pacific. The St.

Anthony Dune Tiger Beetle is found on a few isolated dune fields in southeastern Idaho, of which the St. Anthony Dunes are the most extensive. Recently an isolated population was located east of the Continental Divide in adjacent Montana (Beaverhead County). This population occurs on small hills of loose sand among sagebrush.

Map 24 St. Anthony Dune Tiger Beetle, *Cicindela arenicola*.

Behavior: Adults are active when the temperature of the surface is between 19° and 45° C. Adults mate primarily after they emerge in the spring.

Seasonality: Adults are active from mid-April to late June and then again from late August to late October, although a few individuals can be found throughout the summer and into early November in warm years. In low rainfall years, the life cycle from egg to adult may take up to 4 years.

Larval biology: Burrows occur over much of the dune area but are most concentrated on the flat, grassy interdunal swales adjacent to the dune slopes. Larvae plug the burrow opening when the sand surface becomes too warm and dry.

Bruneau Dune Tiger Beetle, *Cicindela waynei* Leffler
(Plate 6) [Map 25]

Description and similar species: Length 11–13 mm; greenish above. The maculations are very broad and so confluent that the individual maculations are obscured. Extremely similar to the St. Anthony Dune Tiger Beetle in general appearance, behavior and habitat, the Bruneau Dune Tiger Beetle was, until recently, considered a geographical variant of the St. Anthony Dune Tiger Beetle. However, the green upperparts, genitalic differences and the presence of a distinctive tooth that projects up on the mandible (Fig. 4.28) rather than being in line with the rest of the teeth distinguish it.

Subspecies and morphological variants: Possible intermediate forms are found in small sand dunes of south central Idaho (Minidoka County), but none of these individuals exhibit the peculiar mandibular tooth typical of the Bruneau Dune Tiger Beetle.

Map 25 Bruneau Dune Tiger Beetle, *Cicindela waynei.*

Distribution and habitats: As in other sand dune species, the Bruneau Dune Tiger Beetle is most active in the sparsely vegetated swales separating active dunes. Individuals occasionally fly up onto the dune face, but usually make their way back down to the swale area within a few minutes. Endemic to the Bruneau Sand Dunes area of southwestern Idaho.

Behavior: No other species of tiger beetle shares these sand dunes with the Bruneau Dune Tiger Beetle. The adults mate and oviposit primarily on the leeward base of dunes.

Seasonality: Adults have been found from early April to mid-May and then again from mid-July to early October.

Larval biology: The larva is undescribed but probably shares a biology similar to that of the Coral Pink Sand Dune Tiger Beetle and the closely related St. Anthony Dune Tiger Beetle.

Columbia River Tiger Beetle, *Cicindela columbica* Hatch (Plate 6) [Map 26]

Description and similar species: Length 12–13 mm; dark brown above with bold but separated maculations. The front maculation is shaped like an open "C", and the middle line has a steep curve backward with only the hint of a sharp "elbow". The three maculations are not connected along the outer edge of the elytra. Most easily confused with the Western Tiger Beetle, Bronzed Tiger Beetle, and Hairy-necked Tiger Beetle, all of which occur in the same habitat, the Columbia River Tiger Beetle is distinguished by the lack of a sharp "elbow" on the middle mark, heavier maculations in general, darker brown above, and lack of thick hair-like setae on side of thorax.

Subspecies and morphological variants: No geographical variants are known.

Distribution and habitats: Historically found only on sandy beaches and sand bars along the Snake, Salmon and Columbia Rivers of eastern Washington and Oregon and adjacent Idaho, this species at one time occurred from the Dalles in Oregon east to Lewiston, Idaho. Because of dam construction and flooding of habitat for irrigation and hydroelectric projects, most of the habitat of the Columbia River Tiger Beetle has

Map 26 Columbia River Tiger Beetle, *Cicindela columbica*.

been destroyed. It is now known from only a few sites along the Salmon River in Idaho.

Behavior: Adults may occur in large numbers in appropriate habitat. Other species found regularly on these sand bars and river beaches are the Western Tiger Beetle, Bronzed Tiger Beetle, Hairy-necked Tiger Beetle, Wetsalts Tiger Beetle, and Oblique-lined Tiger Beetle. The Columbia River Tiger Beetle is quite wary and has strong escape flights.

Seasonality: Adults are active from mid-April to late June and then again in early August to late September.

Larval biology: The larva is undescribed.

Pacific Coast Tiger Beetle, *Cicindela bellissima* Leng
(Plate 6) [Map 27]

Description and similar species: Length 12–13 mm; above brown with coppery reflections. The maculations are heavy but separated into three distinct lines. The middle line has a curved bend toward the rear but lacks the sharp "elbow" of other species, such as the Western Tiger Beetle, with which it occasionally occurs. Extremely similar in appearance to the Columbia River Tiger Beetle, they are most easily distinguished by their separate geographical ranges.

Subspecies and morphological variants: The brown upperparts of the thorax and head are more greenish in southern Oregon. Individuals that have completely green or blue above are present in small numbers at most locations. The population in the isolated area of Neah Bay in the extreme northwestern Olympic Peninsula of Washington show consistently smaller body size and thinner maculations, and they are recognized by some as a subspecies *C. bellissima frechini* Leffler.

Distribution and habitats: Occurring only in coastal Washington to extreme northern California, this species is restricted to a narrow band of discontinuous sand dunes. The adults are most common in the dune swales and actively shifting dunes from the shoreline inland up to 2 km. Only dunes

Map 27 Pacific Coast Tiger Beetle, *Cicindela bellissima*; **A**, *C. b. bellissima*; **B**, *C. b. frechini*.

that are lightly vegetated or devoid of vegetation are occupied. Where the dunes are relatively close to the ocean shoreline, adults can also be found along the water's edge. Growth of extensive vegetation and planting for dune stabilization produces unsuitable habitat for the Pacific Coast Tiger Beetle.

Behavior: Most adults are solitary, but occasionally aggregations of adults accumulate on small, isolated plains of well-packed sand within the dune field. Adult activity is limited to sand surface temperatures from 30 to 36° C (air temperature 24 to 27° C). Apparently because of the persistent winds in this habitat, the Pacific Coast Tiger Beetle spends considerable time motionless and often digs burrows in which to escape the heaviest winds or on cloudy and cool days. The only other tiger beetle species with which the Pacific Coast Tiger Beetle regularly occurs is the Western Tiger Beetle.

Seasonality: Although apparently a spring–fall species, adults have been observed from early April to early September with most records from May to August.

Larval biology: The larva has been described, but little is known of its biology.

Big Sand Tiger Beetle, *Cicindela formosa* Say
(Plate 7) [Map 28]

Description and similar species: Length 14–21 mm; one of the largest North American species in the tribe Cicindelini, the Big Sand Tiger Beetle is dark above with the same color on the head, thorax and elytra. However, this color varies geographically from dark brown to bright coppery red, dark red, purple, greenish red, and rarely bright green. The ivory-white maculations on the elytra also vary geographically from almost completely covering the elytra to almost completely absent. The metallic color below varies geographically from dark green to blue or purple. Although the Big Sand Tiger Beetle occurs together with many other sand-dwelling species, it is most consistently found with the Festive Tiger Beetle throughout its range. The Festive Tiger Beetle is a smaller species with maculations reduced to a few spots or a thin band around the edge of the elytra. In the

Great Plains, the smaller but similarly colored Blowout Tiger Beetle occupies the same habitat and can be confused with the Big Sand Tiger Beetle. However, the Big Sand Tiger Beetle is bulkier and usually larger with a shorter labrum.

Subspecies and morphological variants: At present this species includes five recognized subspecies, but several additional forms are likely to be distinguished with more studies. Perhaps because *C. formosa*, especially in the west, is restricted to often isolated sandy areas, and because it is apparently not good at dispersing, the evolution of adaptations in coloring for local conditions has led to distinct populations. Thermoregulation and camouflage appear to be the two main adaptations affecting color patterns of each population.

C. formosa formosa Say: Bright coppery-red above with very broad maculations that are connected to each other with coalescence of maculations on some individuals so extreme that the first and rear maculations become obscured. It is metallic purple below. This nominate form occurs west of the Missouri River to the eastern base of the Rocky Mountains. Intergrades with *C. f. generosa* occur in a broad zone on either side of the Missouri River. Narrow zones of intergradation occur with *pigmentosignata* in east Texas and with *rutilovirescens* along the Texas-New Mexico border. Individuals that are bright green above occur with the more common red forms in northeastern Wyoming.

C. formosa generosa Dejean: Dark brown above with wide but distinct maculations. They are joined together by a thin band along the outer edge of the elytra. It is metallic dark green below with some coppery reflections. This subspecies occurs over most of the northern United States east of the Missouri River but is absent from the southeastern states. A significant number of individuals in northwestern Minnesota, northeastern North Dakota and adjacent Manitoba have greatly expanded ivory maculations and were at one time considered a separate subspecies, "manitoba." Current research, however, indicates that this population probably is not consistent enough to be considered a separate subspecies. An isolated population with maculations intermediate between *C. f. gibsoni* and *C. f. formosa* has been discovered recently in southwestern Montana. Intergrades are known between *generosa* and the nominate race in a broad zone along the Missouri River.

C. formosa gibsoni Brown: Ranges from dark red to purple above with the maculations so expanded and coalesced in most individuals that they cover the elytra completely except for a broad dark wedge down the length of the middle. On some individuals, however, a small dark area expands behind the middle line to extend this dark wedge to the rear. It is metallic bluish-violet below. The subspecies *gibsoni* is restricted to two areas separated by 1100 km. In and immediately around the Great Sand Hills of southwestern Saskatchewan it intergrades narrowly with nominate *formosa* on all sides of this sand dune area. In the second area, the population of *gibsoni* in similar habitat in the Maybell Sand Dunes of Moffat County, in northwestern Colorado, is isolated to the east by 230 km from other populations of *formosa*. To the west it intergrades in a narrow zone with nominate forms along the Green River in northwestern Utah. The two separated populations of *gibsoni* show little difference except in the head capsule coloring of the third instar larva. This general similarity may best be explained as an independent evolution of coloring to adapt to similar white sand condition in the two areas. If so, this convergence onto an indistinguishable pattern by coincidence would then not strictly meet the definition of a single subspecies designation for the two populations.

C. formosa pigmentosignata **Horn**: This beautiful subspecies has dark upperparts violaceous with deep purple. The maculations range from absent except for a thin ivory line on the edge of the end of each elytron to complete maculations connected by a thin line along the side edge of each elytron. Metallic bluish-green below. This subspecies is restricted to sandy areas of open pine forest from where the corners of Texas, Arkansas and Louisiana come together southwest into central Texas. It intergrades in a narrow band with nominate populations to the northwest but is isolated by 300 km from the nearest known populations of *generosa* in southeastern Louisiana and northeastern Arkansas by the Mississippi River flood plain.

C. formosa rutilovirescens **Rumpp**: The upperparts of the head and thorax of this subspecies are reddish-green to reddish-purple. The elytra almost completely lack ivory-white maculations except for a thin line along the rear edge of the elytra. Deep metallic violaceous blue below. This population is restricted to low dunes of the Mescalero Sands interspersed with short scrub oak above 1000 m elevation. This dune field is limited to east central New Mexico and adjacent Texas. Intergradation occurs with the nominate subspecies at the edges of this dune area to the east and north.

Distribution and habitats: Everywhere it occurs, the Big Sand Tiger Beetle occupies dry upland sandy areas with little or no low vegetation and no standing water. In the east it is found in road cuts, sandy fields, sea shore dunes, and pine barrens. In the west it is more restricted to extensive sandy blowouts and dunes. The species occupies most of the United States and extreme southern Prairie Provinces of Canada east of the Rocky Mountains. However, it is conspicuously absent from southern and western Texas, the southern

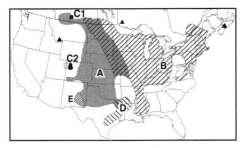

Map 28 Big Sand Tiger Beetle, *Cicindela formosa*; **A**, *C. f. formosa*; **B**, *C. f. generosa*; **C1** and **C2**, *C. f. gibsoni*; **D**, *C. f. pigmentosignata*; **E**, *C. f. rutilovirescens*.

Mississippi flood plain, the Gulf of Mexico flood plain and the southeastern United States even where there is what appears to be appropriate habitat. Throughout most of its range, the Big Sand Tiger Beetle commonly occurs together with the Festive Tiger Beetle.

Behavior: Because of its large size, the Big Sand Tiger Beetle apparently can prey on items as large as small tiger beetle species. It is an important predator on the Ghost Tiger Beetle where their ranges overlap. Also because of its large size, it warms up more slowly than its smaller neighbors and becomes active later in the morning. Where sandy areas become vegetated and stabilized, its populations tend to decline. To escape danger, this species flies long distances and often tumbles when it lands.

Seasonality: In the southern part of its range, adults are active from April to July and August to October. In northern parts of its range, late spring and early fall activity patterns often lead to a few adults found throughout the summer. The life cycle is 2 or more years depending on latitude and food availability. Both adults and larvae overwinter.

Larval biology: Burrows are found in open areas of well-drained soil with no or sparse vegetation. Larvae of this species dig the deepest burrows known for tiger beetles (30–200 cm). Apparently the great depth of their burrows allows larvae to survive the winter below the frost line, and the deepest burrows are at the highest latitudes. A unique cup-like pit next to the opening at the surface apparently aids in capturing prey.

Cow Path Tiger Beetle, *Cicindela purpurea* Olivier
(Plate 8) [Map 29]

Description and similar species: Length 12–16 mm; above dark red and purple to greenish-red or lime green with reddish edges. In some populations all black individuals occur. The maculations on the elytra are variable but reduced in most populations to a short middle line that is angled backwards and isolated from any other maculations. The rear maculation is a thin line along the rear tip of the elytron and occasionally has an isolated dot just in front of it. Some populations have no maculations whatsoever on the elytra, and at least one population has a heavy white line along the side of the elytra that connects the front, middle and rear maculations. The legs are usually the same color as the upperparts, but in some populations they are coppery in contrast to the green upperparts. Below metallic and varies in color from dark green to bluish. The similar Green Claybank Tiger Beetle has two dots that make up the front maculation; the middle maculation is perpendicular to the outer edge of the elytra and has a definite bend backwards. The middle line of the Splendid Tiger Beetle is usually a short horizontal line, and the head and thorax are green or blue in contrast to the reddish elytra.

Subspecies and morphological variants: More than fifteen forms of this widespread and variable species have been described. Several have since been recognized as separate species, and others have been determined to be invalid or duplicated names. At present five subspecies of the Cow Path Tiger Beetle are recognized.

C. purpurea purpurea Olivier: This nominate subspecies is purplish with green tinges above. The maculations are reduced to a narrow middle line that does not reach the outer edge of the elytron. It occurs east of the Mississippi River and intergrades with *audubonii* in a wide zone from the Dakotas south to Oklahoma.

C. purpurea audubonii LeConte: Above generally greenish, often purplish tinged and a distinct metallic purple border around the edge of the elytra. The maculations are reduced to a single angled middle line and a white rear tip at the edge of the elytra, occasionally with a small white spot just above it. Individuals that are black above and below but with the same maculations are common in some areas, especially in the northern and western parts of its range. Here black individuals can often make up 20 to 40% of the population. Intergrades with the nominate subspecies occur in a broad zone along the Red River of the North and the lower Missouri River.

C. purpurea cimarrona LeConte: Above, this subspecies is muddy green to blackish or dark brown often with reddish coloring. Most distinct, however, are the heavy maculations that are usually completely or partially connected by a white band running along the outer edge of each elytron. Occurring from central Colorado to southern Arizona and New Mexico, this subspecies is generally found in grassy meadows at higher elevations, especially in the southern parts of its range.

C. purpurea hatchi Leffler: This subspecies is grassy green above with coppery legs and distinct coppery patches on the head and sides of elytra. Restricted to open coniferous forest of the foothills of the central Sierra Nevada Mountains in northern California and the western foothills of the Willamette Valley in Oregon north to southern Vancouver island, British

Columbia, it intergrades with *lauta* in the Willamette Valley of Oregon south to Shasta and Modoc counties in northwestern California.

C. purpurea lauta Casey: This subspecies is grassy green above with indistinct coppery patches on the head and thorax, and the legs are green. It occurs from northwestern California through the Klamath Mountains and north along the eastern foothills of the Willamette Valley to the Columbia River.

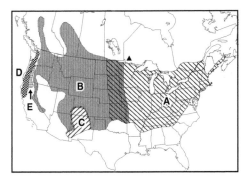

Distribution and habitats: Found from the Atlantic to the Pacific, the Cow Path Tiger Beetle occurs away from standing water in open fields, grasslands, scrubby forest clearings, and in mountain meadows at higher elevations in the west. Often even small patches of open grasslands are sufficient to support populations of this species. It is absent from the southeastern United States and desert southwest.

Map 29 Cow Path Tiger Beetle, *Cicindela purpurea*; **A,** *C. p. purpurea*; **B,** *C. p. audubonii*; **C,** *C. p. cimarrona*; **D,** *C. p. hatchi*; **E,** *C. p. lauta*.

Behavior: Individuals are often found at low densities, usually on bare patches of soil. They are quick to take flight and often land in grassy vegetation. It is most commonly found together with species such as the Long-lipped Tiger Beetle, Oblique-lined Tiger Beetle, and the Variable Tiger Beetle.

Seasonality: Adults are one of the earliest species to appear in the spring, and are active from February to early June and late August to October. The life cycle is 2–3 years, and both adults and larvae overwinter.

Larval biology: Burrows are scattered in small bare patches of clay to sandy-clay soils. They are shallow (8–15 cm) and probably limited in depth by the hard, compact soil in which the species usually occurs.

Ohlone Tiger Beetle, *Cicindela ohlone* Freitag and Kavanaugh (Plate 8) [Map 30]

Description and similar species: Length 10–13 mm; only discovered and described in the 1990s, this species is grassy-green above with subtle bronze highlights on the thorax and around the middle maculation on the elytra. The middle maculation is heavy with a distinct curve rearward. The front maculation is composed of two isolated dots, one hidden under the shoulder and the other more obviously situated on the upper surface of the elytron. The rear maculation is a short, bold line at the tip of the elytra with a dot above it. The legs are green. This species is most similar to *C. purpurea lauta* but is recognized as a separate species because of genitalic differences, smaller size, more rounded outer edges of elytra, and different seasonal activity patterns.

Subspecies and morphological variants: No geographical variants are known for this highly endemic species.

Map 30 Ohlone Tiger Beetle, *Cicindela ohlone*.

Distribution and habitats: Known from several sites in Santa Cruz County, California, this species occurs in native coastal terrace grasslands at the southwestern base of the Santa Cruz Mountains between 60 and 340 m elevation. No other species of tiger beetle occurs in this habitat. **NOTE:** Because of its small range and the likelihood of human destruction of its remnant habitat and possibility of extinction, this species was listed as an ENDANGERED species in October 2001 by the US Fish and Wildlife Service. Collection of specimens is illegal.

Behavior: Adults are most often found along trails and in bare patches of low grass. They fly to denser grass when disturbed. Daily activity is sporadic because of frequent periods of cloudy and rainy weather. At these times adults seek shelter in the bases of grass clumps.

Seasonality: This species has an unusual winter–spring seasonal activity. Adults are active from late January to early April and no fall records are known. The life cycle is 1–2 years.

Larval biology: Larvae occur with the adults along the edges of paths and scattered bare patches among the grass. The extremely hard and compact clay soil limits burrow depth to 8–15 cm.

Sagebrush Tiger Beetle, *Cicindela pugetana* Casey
(Plate 8) [Map 31]

Description and similar species: Length 13–15 mm; above either bright green, blue green or black. The maculations are reduced to a thin white line on the rear tip of each elytron and, on some individuals, a thin middle line that is not connected to the outer edge of the elytra. The underparts and legs are the same color as the upperparts. The upper lip (labrum) on females is long and black

95

but shorter and ivory-white on males, a character shared by the very similar but higher elevation species, the Alpine Tiger Beetle. Apart from their habitats, these two species are distinguished with difficulty by subtle differences in the shape and structure of the labrum and shinier surface on the elytra of the Alpine Tiger Beetle. The only other green tiger beetle to share the sagebrush habitats of the Northern Sagebrush Tiger Beetle is the Badlands Tiger Beetle, which has bold middle and rear maculations, and both the male and female have ivory-white labra.

Subspecies and morphological variants: At present no distinct subpopulations of the Sagebrush Tiger Beetle are known, however, with further studies, this species may prove to be a form of the Alpine Tiger Beetle.

Map 31 Sagebrush Tiger Beetle, *Cicindela pugetana.*

Distribution and habitats: Restricted to sagebrush areas of the northern Great Basin from extreme southern British Columbia south to north central Oregon, this species occupies bare soil patches and road cuts through low brush.

Behavior: Although not gregarious, individuals, often both black and green forms together, appear to forage in patches of several hectares and then be absent for great distances from intermediate areas of what appear to be similar habitat.

Seasonality: Adults are active from early March to mid-June and then again for a short period in late September to late October.

Larval biology: Larva unknown.

Alpine Tiger Beetle, *Cicindela plutonica* Casey
(Plate 8) [Map 32]

Description and similar species: Length 13–16 mm; above shiny green, blue-green, or black. There are usually no maculations, or if present, they are reduced to an almost imperceptible white dash on the tip of each elytron. The underparts and legs are the same color as the upperparts. Females of all color forms have a long black upper lip (labrum), but that of the male is shorter and ivory white. The only other tiger beetle species likely to be encountered at high elevations in northeastern California and adjacent Oregon is the

Dispirited Tiger Beetle, which is duller, has more complete maculations and a broader body shape.

Subspecies and morphological variants: The green form of this species was originally described as the subspecies **C. p. *leachi* Cazier,** but because both black and green forms occur so regularly together at the same sites, commonly copulate with each other, and show no distinct differences in genitalia, they are better considered color morphs of the same species. Because of similarities in body structures, habitat use and behavior, some experts suggest that the Sagebrush Tiger Beetle may be a subspecies of the Alpine Tiger Beetle.

Map 32 Alpine Tiger Beetle, *Cicindela plutonica.*

Distribution and habitats: At higher elevations up to 2700 m this rare species occurs in bare rock granite hillsides with patches of melting snow. At lower elevations it appears in bare patches of soil in sagebrush and high desert areas.

Behavior: The pattern of observations for this species indicates it occurs in small colonies covering less than a hectare. It is then absent from intervening areas for considerable distances.

Seasonality: At higher elevations, adult are active from early June to early July, however, deep snow earlier in the season makes accurate observation of this species difficult. A fall activity period has not been observed. At lower elevations in Oregon, Idaho, and Utah, adults are active as early as February or March and again in September and October.

Larval biology: Larva unknown.

Splendid Tiger Beetle, *Cicindela splendida* Hentz
(Plate 9) [Map 33]

Description and similar species: Length 12–15 mm; distinguished by the contrasting colors above. The head and thorax are metallic green to blue and the elytra are brick red. The maculations are variably reduced to thin disconnected lines and dots or partially absent. Other similar reddish species, such as the Green Claybank Tiger Beetle and the Cow Path Tiger Beetle, bear only a single color above.

Subspecies and morphological variants: Some populations have heavier maculations, but no population is consistently different enough to justify a status of subspecies. However, many individuals in northern populations have strong maculations resembling those of the Common Claybank Tiger Beetle. In northwestern Louisiana and southwestern Arkansas west to the Dallas area of Texas individuals with green elytra and blue head and thorax occur ("ludoviciana"). These forms are in the same habitats as and copulate with the normal Splendid Tiger Beetle, but show no genitalic differences. Some experts consider them an isolated population of the all green western species Green Claybank Tiger Beetle, but the ecology, behavior and distribution of these greenish "ludoviciana" forms leads us to consider them more likely to be a local color variant of the Splendid Tiger Beetle. However, these distinctions may be moot as recent DNA studies show the Splendid Tiger Beetle is so similar genetically to the Green Claybank Tiger Beetle and the Common Claybank Tiger Beetle that they may be a single species.

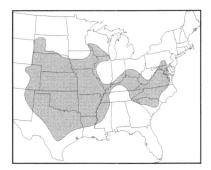

Map 33 Splendid Tiger Beetle, *Cicindela splendida*.

Distribution and habitats: This species is associated with open red clay soils, including road cuts, banks, dirt roads, and areas with sparse vegetation. The distribution of this species is tightly associated with the patchy distribution of these clay soils.

Behavior: Even in its localized colonies, the Splendid Tiger Beetle is rarely gregarious. Adults are quick to take flight, and these escape flights are long. It is commonly found together with the Cow Path Tiger Beetle and Common Claybank Tiger Beetle.

Seasonality: On south-facing slopes, adults can be active as early as late January into April and then again from September to October. The life cycle is 2–3 years.

Larval biology: Burrows are found in red clay soils, especially on slopes, road cuts and banks. Burrow depth is 10–15 cm.

Green Claybank Tiger Beetle, *Cicindela denverensis* Casey
(Plate 9) [Map 34]

Description and similar species: Length 12–15 mm; all green to blue-green above, this species has variable but thin and reduced maculations. It could be confused with the Six-spotted Tiger Beetle, but the Green Claybank Tiger Beetle occurs in grassy habitats of the western prairies and not hardwood forest of the east. The Green Claybank Tiger Beetle usually has maculations that are thin, short lines, not spots.

Subspecies and morphological variants: No subspecies or local variants are known. The form "ludoviciana" from northwestern Louisiana and southwestern Arkansas to northeastern Texas may be either an isolated population of the Green Claybank Tiger Beetle or a local green morph of the Splendid Tiger Beetle. Individuals from southwestern Nebraska tend to have strong maculations resembling those of the Common Claybank Tiger Beetle. However, recent DNA studies show the Green Claybank Tiger Beetle is so similar genetically to the Splendid Tiger Beetle and Common Claybank Tiger Beetle that they may be a single species.

Map 34 Green Claybank Tiger Beetle, *Cicindela denverensis.*

Distribution and habitats: Widespread in the western Great Plains but patchy distribution in grassy areas and prairies, especially in dry gullies and near clay banks.

Behavior: Encountered in small numbers, it is somewhat gregarious and a strong flier. Commonly occurs together with the Splendid Tiger Beetle and the Festive Tiger Beetle.

Seasonality: Adults active from March to June and from July to early November, but in exceptionally early spring seasons they have been recorded active as early as late February in southeastern Colorado.

Larval biology: Larva unknown.

Common Claybank Tiger Beetle, *Cicindela limbalis* Klug
(Plate 9) [Map 35]

Description and similar species: Length 11–16 mm; bright reddish to reddish-green above with distinct maculations. The middle maculation is quite variable but usually horizontal with a wave-like bend in its middle that generally reaches all the way to the outer edge of the elytra. Some individuals, however, have very small middle maculations. Metallic copper and green below. It is most similar to reddish forms of the Cow Path Tiger Beetle but is usually distinguished by the middle line on the elytra. In the Cow Path Tiger Beetle this line is short and does not reach the edge of the elytra. Purplish forms of the Badlands Tiger Beetle are also quite similar, but the middle line does not reach the edge of the elytra.

Subspecies and morphological variants: No distinct geographical forms are recognized by most taxonomists, but some individuals from the east slope of the Rocky Mountains in Colorado have their maculations broadly connected along the outer edge of the elytra. The most completely maculated forms of this highly variable population have been described as the subspecies *C. l. sedalia* **Smyth**. Recent DNA studies show that the Common Claybank Tiger Beetle is so similar genetically to the Splendid Tiger Beetle and the Green Claybank Tiger Beetle that they may be a single species.

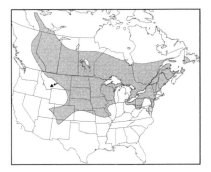

Distribution and habitats: Widely distributed and usually associated with red clay or bare soil patches with sparse vegetation on banks, roads, forest openings and bare slopes. Apparently extirpated from most of southern New England.

Map 35 Common Claybank Tiger Beetle, *Cicindela limbalis.*

Behavior: Solitary but occurs regularly together with the Cow Path Tiger Beetle and the Splendid Tiger Beetle.

Seasonality: Adults are active from March to June and from July to September.

Larval biology: Burrows most common on sloping, often steep, clay soils. Burrow depths 7–15 cm and occasionally with a short chimney-like extension above the opening.

Badlands Tiger Beetle, *Cicindela decemnotata* Say
(Plate 9) [Map 36]

Description and similar species: Length 12–14 mm; green to rarely reddish above with maculations generally bold but almost absent in some individuals. If present, the middle maculation is a line curving rearward at a sharp angle and does not reach the outer edge of the elytra. Metallic blue-green below. Maculations suggest those of the Green Claybank Tiger Beetle, but the mid line of the Green Claybank Tiger Beetle is bent less acutely and extends to the edge of the elytra. Also, the underside of the Green Claybank Tiger Beetle is green with distinctive coppery tones on the underside of the thorax. In the Badlands Tiger Beetle the entire undersides are metallic green to blue-green. Similar to green forms of the Sagebrush

Tiger Beetle, the Badlands Tiger Beetle is distinguished by a white labrum on both males and females.

Subspecies and morphological variants: Recently four geographical populations have been described as separate subspecies based on their DNA and body color patterns.

C. decemnotata decemnotata **Say:** This population exhibits the full range of colorations from green, dark green, blue-green, olive, green-purple, purple, and red-purple to black. The maculations on the elytra are generally complete. This is the most widely distributed subspecies, and it occurs from east central Alaska south to northern New Mexico and west into southern Utah and Idaho.

C. decemnotata meriwether **Knisley and Kippenhan:** Restricted to the Columbia Basin of central Washington and southern British Columbia, this population is recognized by the combination of the back green with narrow side margins of the elytra bright green, dark green, or blue-green. The maculations on the elytra are reduced to a middle band, two dots at the end of the elytra, and occasionally a dot at the shoulder.

C. decemnotata bonnevillensis **Knisley and Kippenhan:** Found only on the western border of ancient Lake Bonneville in the western Utah desert of Tooele County, this subspecies occupies more saline soils than any of the other subspecies. Its back is bright green to dark blue or purple with greatly reduced elytral maculations showing small vestiges of the middle band to virtually no maculations. Individuals superficially resemble the Alpine Tiger Beetle.

C. decemnotata montevolans **Knisley and Kippenhan:** This subspecies is distinguished by a combination of a dull red-purple back, elytra with no contrasting side margin, and the elytral maculations reduced to a few dots. It occurs at elevations above 2000 m in the Bear River Mountain Range of southeastern Idaho and northeastern Utah.

Distribution and habitats: Found in patchy colonies in a variety of habitats including sparsely vegetated grasslands, open pine forests, sage brush, saline flats, chalky slopes and open brushy areas with clay, sandy, or gravel soils.

Behavior: This is a solitary species with individuals widely spaced in its open habitat. It is a strong flier and not attracted to night lights. Found together with the Cow Path Tiger Beetle, and Long-lipped Tiger Beetle.

Map 36 Badlands Tiger Beetle, *Cicindela decemnotata;*
A, *C d. decemnotata;*
B, *C. d. bonnevillensis;*
C, *C. d. meriwetheri;* **D,**
C. d. montevolans.

Seasonality: Adults active from mid-March to late June and again in early September to mid-October. Higher elevation populations are active from late June to late August.

Larval biology: Larvae take 2 years to develop into adults, but in higher elevations and latitudes they may take 3 years.

Six-spotted Tiger Beetle, *Cicindela sexguttata* Fabricius
(Plate 9) [Map 37]

Description and similar species: Length 10–14 mm; bright metallic green to blue-green above with maculations reduced to six, four, or two small spots. In the hand, the rear tip of the elytra has small but distinct sawtooth edges. Below the same color as above. This common species can be confused with several other metallic green species with reduced maculations. However, the western range and grassland habitat of several of these, such as the Green Claybank Tiger Beetle and the Cochise Tiger Beetle, are so different they do not overlap with the Six-spotted Tiger Beetle. Several green subspecies of the Festive Tiger Beetle do occur in the same area and habitat, but they have stouter bodies with more rounded elytra, and the females have dark upper lips (labia). The green form of the Northern Barrens Tiger Beetle often occurs in the same habitat as the Six-spotted Tiger Beetle, but the Northern Barrens Tiger Beetle can be distinguished by its duller color and a complete middle line on the elytra. The closely related and extremely similar Laurentian Tiger Beetle is separated by larger size, genitalic differences, and nonoverlapping range in boreal forest.

Subspecies and morphological variants: At least ten subspecies of the Six-spotted Tiger Beetle have been proposed, but current taxonomic analysis shows that none of them is consistently different from other populations. In Nova Scotia, a small proportion of individuals (mainly males) in some populations are black (melanistic). In Nebraska many individuals are purplish.

Distribution and habitats: A species of loamy to sandy soil in eastern hardwood forests and occasionally mixed open pine forest, it is solitary and found most often along trails and in sun flecks on the forest floor as well as the forest edge.

Map 37 Six-spotted Tiger Beetle, *Cicindela sexguttata*.

Behavior: Normally solitary, individuals warm themselves in sun flecks on the forest floor. As summer advances and leaves shade the forest floor more and more, small groups of this tiger beetle will sometimes gather at the few sun flecks still reaching the ground. This is the common green species seen by hikers and others walking along woodland trails.

Seasonality: Adults are primarily active in the spring from April to early July. In some areas, however, they can be found regularly in the fall until September. This is one of the few species in North America exhibiting a dominantly spring adult activity cycle.

Larval biology: Burrows are along woodland trails or forest edges in low densities, often near logs or fallen branches on the forest floor.

Laurentian Tiger Beetle, *Cicindela denikei* Brown (Plate 9) [Map 38]

Description and similar species: Length 13–15 mm; bright metallic green to blue-green above. Maculations reduced to small spots. Closely related and so similar to the Six-spotted Tiger Beetle, the Laurentian Tiger Beetle is called a cryptic or hidden species. It is separated by its larger size and occurrence in exposed soil habitats that are sandy and gravely. The two species do not overlap in distribution.

Subspecies and morphological variants: No geographically distinct populations are known for this restricted-range species.

Map 38 Laurentian Tiger Beetle, *Cicindela denikei.*

Distribution and habitats: The Laurentian Tiger Beetle is highly adapted to open patches of sparse vegetation called alvars. These patches within the Boreal forest are formed over shallow exposed bedrock (granite or sandstone) and have thin or no soil. Because of this poor substrate, trees and shrubs are stunted, grasses are common, and alvars resemble prairie habitats. They are also strongly affected by spring flooding and mid-summer drought. Most alvars in North America are in the Great Lakes Basin, and only 120 of them exist with a total area of 112 km². The Laurentian Tiger Beetle until recently was thought to be restricted to alvars northeast of Lake Superior, but recently a population was found on Manitoulin Island between Lake Huron and the Georgian Bay in southeastern Ontario.

Behavior: In addition to typical active foraging, this species often hunts for prey by waiting in ambush and then pouncing on it. Usually few other tiger beetle species occur in this habitat.

Seasonality: Adults active from May to August but most common in June and July.

Larval biology: Burrows are found in sandy and gravel soils and usually under a flat rock. Depth of burrows 10–20 cm.

Northern Barrens Tiger Beetle, *Cicindela patruela* Dejean
(Plate 9) [Map 39]

Description and similar species: Length 12–14 mm; dull metallic green, greenish-brown, or black above. Metallic green or black below. Three distinct maculations are present but disconnected, and the middle line, which is complete, reaches to the outer edge of the elytra. Green forms of the Northern Barrens Tiger Beetle occur with and closely resemble the Six-spotted Tiger Beetle. They are readily distinguished by the duller green color and complete middle band on the elytra of the Northern Barrens Tiger Beetle.

Subspecies and morphological variants: Based on color two subspecies have been described.

    C. patruela patruela **Dejean:** This green form occurs throughout the majority of the species' range. In Monroe County of southwestern Wisconsin, a small proportion of individuals have the green upperparts replaced with muddy green, brown or black. These have been described as a separate subspecies, "huberi", but more recent studies suggest they are better considered a local color variant.

    C. patruela consentanea **Dejean:** In the pine barrens of Long Island, New Jersey and Delaware, populations are identical to nominate forms except that the green upperparts and underparts are replaced by black.

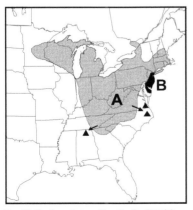

Map 39 Northern Barrens Tiger Beetle, *Cicindela patruela*; **A**, *C. p. patruela*; **B**, *C. p. consentanea*.

Distribution and habitats: Although this species has a wide range across the northeastern and northern Great Lakes area of the United States, its populations are discontinuous and localized. It often occurs in dry sandy soils of mixed oak and pine forest usually in clearings of eroded sandstone and often associated with mosses, lichens and ground-hugging vegetation. In parts of its range it may be found in other poor soils besides sand, such as eroding shale or exposed crumbling rock on hillsides, but again usually on dry slopes with lichens and dry mosses. In the southeastern parts of its range, it is found only in higher elevations above 500 m.

Behavior: The light sandy substrate on which the Northern Barrens Tiger Beetle occurs can make this species, especially the green forms, obvious. However, when they fly to escape danger, green patches of moss or low vegetation are usually the target landing area where they are better camouflaged. The black *consentanea* form resembles bits of forest floor litter and pieces of charcoal that are often present in the fire-prone pine barrens of New Jersey and New York, and the brownish "huberi" form matches dead or "bleached" lichens common on the forest floor within its range. The Northern Barrens Tiger Beetle is regularly found together with Six-spotted Tiger Beetle, Festive Tiger Beetle, and farther north with the Long-lipped Tiger Beetle.

Seasonality: Adults are active from April to July and then again from August to October, although fall activity in some areas is reduced or absent in some years. The life cycle is 2 years.

Larval biology: Burrows occur in scattered open patches of stabilized and compact sandy soils, typically in low densities.

Beautiful Tiger Beetle, *Cicindela pulchra* Say (Plate 10) [Map 40]

Description and similar species: Length 15–18 mm; above brilliant copper red to dark red with metallic dark green, purple or blue borders on the elytra, thorax and head; below metallic green to blue. The maculations are reduced. The middle line is a short wedge on the outer elytral edge, and, if present, the front and rear maculations are small dots also on the edge of the elytra, although in southern populations these three marks are often connected by a white line along the outer edge of the elytra. This species can only be confused with nominate Festive Tiger Beetle, which is much smaller and has contrasting blue or green head and thorax with reddish-orange elytra.

Subspecies and morphological variants: Two subspecies are recognized, and they differ in the brightness of the reddish elytra and the extent of the maculations.

C. p. pulchra **Say**: Found in the Great Plains it typically has reduced maculations and is mainly dark red above.

C. p. dorothea **Rumpp**: Occurs from southeastern Arizona to west Texas. It is smaller, has wider maculations that are often connected to each other along the outer elytral edge, and above it is more metallic red to red-green.

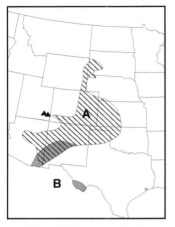

Map 40 Beautiful Tiger Beetle, *Cicindela pulchra*; **A**, *C. p. pulchra*; **B**, *C. p. dorothea*.

Distribution and habitats: Found in bare soil areas of grasslands and desert saltbush flats of the southwest, this species also occurs in similar habitats in northern Mexico from Sonora and Chihuahua to San Luis Potosi.

Behavior: Solitary and a very strong flier, the Beautiful Tiger Beetle also runs quickly from one bunch grass clump to the next. It often hunts by waiting in the shade of these grass clumps to ambush prey. During the heat of the day, adults seek shelter in the shade of grass clumps. It apparently is a mimic of several species of the similarly colored female velvet ant (Mutillidae) *Dasymutilla,* and even creates vibrations by scraping body parts across one another (stridulates) at a similar sound frequency when captured.

Seasonality: Adult activity periods vary considerably throughout its range. The northern nominate form is active from late February to April and from July to early October. In the southern part of its range the monsoonal rainy season dictates a single period of activity from July to September. The life cycle is 2–3 years.

Larval biology: The shallow (6–15 cm) burrows are found in hard-packed sandy-clay soils in low densities. In some areas the larvae are heavily attacked by the parasitoid tiphiid wasp, *Pterombrus.*

Cochise Tiger Beetle, *Cicindela pimeriana* LeConte
(Plate 10) [Map 41]

Description and similar species: Length 11–14 mm; above and below bright metallic green to blue. The elytral maculations are usually absent, but on a few individuals a short white wedge extends inward from the middle of each elytron edge. A few older individuals darken so that they are almost black.

Subspecies and morphological variants: No geographical variants are known for this highly range-restricted species.

Map 41 Cochise Tiger Beetle, *Cicindela pimeriana*.

Distribution and habitats: Restricted to the Sulphur Springs and San Bernardino Valleys of extreme southeastern Arizona, this endemic species may occur in adjacent areas of New Mexico and Sonora as well. It is limited to white clay banks in open grassy areas, often in the vicinity of standing water. During the summer monsoon season, it also occurs at the edges of temporary puddles and ponds.

Behavior: A solitary species for much of the year, it often congregates in small numbers around pond edges when they are available. It is a strong flier and quite wary.

Seasonality: Often the only upland tiger beetle species active in its range in early spring (April to May) and late fall (October), it is most common following the summer rains in July to September.

Larval biology: The larva has been described, but its biology is unknown.

Crimson Saltflat Tiger Beetle, *Cicindela fulgida* Say
(Plate 10) [Map 42]

Description and similar species: Length 10–13 mm; metallic copper to reddish-brown above. Some populations have green, blue or dark purple individuals. The maculations are heavy and complete. In some populations the maculations coalesce to connect along the outer elytral edge. Metallic green to blue below. Most similar to the closely related Dark Saltflat Tiger Beetle, the two do not occur in the same geographic area and have subtle physical differences such as shorter labrum length in the Crimson Saltflat Tiger Beetle as well as characteristics of the male aedeagus shape.

Subspecies and morphological variants: Although as many as six subspecies of this species have been named, the differences among many geographical populations are often ambiguous and inconsistent. The concept of subspecies for this tiger beetle is weakly supported, and there may be only four populations sufficiently distinct to retain even weak subspecific status. The differences are based mainly on body size and extent and shape of maculations.

    C. fulgida fulgida Say: Occurring in the eastern Great Plains, the nominate subspecies is usually coppery red above but with individuals in some areas purplish to black. The middle maculation is a line bent at a right angle so that it runs parallel to the outer edge of the elytra.

C. fulgida pseudowillistoni W. Horn: Found throughout the western Great Plains and intermontane southern Rocky Mountains, above generally reddish but with considerable variation into brown, blue and green. The middle line is bent at less than a right angle and is oblique to the edge of the elytra. The forms "williamlarsi" and "winonae" are insufficiently distinct from *pseudowillistoni* to be considered subspecies.

C. fulgida rumppi Knudsen: The smallest subspecies, this form is also characterized by very broad maculations that coalesce to cover half or more of the elytral surface. It is restricted to the Laguna del Perro area of Torrance County, in central New Mexico. However, populations in northwestern New Mexico and southeastern Utah that are within the range of *pseudowillistoni* have expanded maculations similar to *rumppi*.

C. fulgida westbournei Calder: Confined to the northern Great Plains and the southern Prairie Provinces, this subspecies has maculations similar to nominate forms but is typically darker above, ranging from purple-red to dark blue and occasionally bright green. The tip of the male aedeagus is also blunter than in other populations.

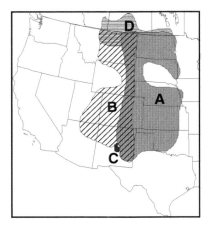

Map 42 Crimson Saltflat Tiger Beetle, *Cicindela fulgida*; **A**, *C. f. fulgida*; **B**, *C. f. pseudowillistoni*; **C**, *C. f. rumppi*; **D**, *C. f. westbournei*.

Distribution and habitats: Occurring throughout the Great Plains and intermontane valleys of the southern Rocky Mountains, this species is confined to moist alkaline soils of salt flats or river and lake edges with short and sparse vegetation. These habitats are associated with lakes, marshes, ponds and rivers where high summer temperatures result in constant evaporation and concentration of high pH (8.5) minerals on the soil surface.

Behavior: A sometimes gregarious species, it usually forages in areas of sparse vegetation and grassy saline areas characterized by a white soil surface. It hides or thermoregulates in the shade of grass clumps and low vegetation. The Crimson Saltflat Tiger Beetle is a relatively weak flier with short escape flights. It often runs into vegetation to avoid danger. At least nine other species of salt flat–occurring tiger beetle species occur with it throughout its range, but none has the bright red color of this species.

Seasonality: Adults are active from April to June and from July to October.

Larval biology: The larva has been described, but its biology is unknown.

Dark Saltflat Tiger Beetle, *Cicindela parowana* Wickham (Plate 10) [Map 43]

Description and similar species: Length 11–12 mm; above dark purple or blue to greenish with strong red reflections. The maculations are very heavy and in some populations expanded so that the first and middle maculations coalesce. Metallic green to blue below. Extremely similar to the Crimson Saltflat Tiger Beetle and evidently closely related to it, the Dark Saltflat Tiger Beetle is best separated by the different geographic range, a longer labrum and subtle differences in the male aedeagus shape. In some areas it may occur together with the similar Williston's Tiger Beetle, which has a much shorter front maculation.

Subspecies and morphological variants: As might be expected for species specialized on a highly localized and widely separated habitat, populations have diverged in appearance, and three subspecies are recognized.

C. parowana parowana Wickham: Found in the Great Basin of eastern Nevada through Utah and north to southwestern Idaho, this subspecies characteristically is dark purple above with copper reflections and very broad but not coalesced maculations. Below it is metallic bluish-green. It intergrades with *wallisi* and *platti* in a narrow zone in eastern Oregon. The subspecies *remittens* Casey has been described for populations in Utah, but we consider it inseparable from the nominate subspecies.

C. parowana platti Cazier: Greenish-red above and maculations so expanded that the front and the middle ones often coalesce; this subspecies is readily separated from other populations. Below it is metallic blue. Occurring from east central California and adjacent Nevada north to Oregon, the southern parts of its range in northwestern Nevada appear to be isolated from southern populations of the nominate subspecies in western Utah and eastern Nevada. In southeastern Oregon, however, *platti* intergrades with *wallisi*.

C. parowana wallisi Calder: Occupying the Great Basin from Oregon north into the Okanogan area of southern British Columbia, where it is now likely extirpated, this subspecies has the narrowest elytral maculations of all the subspecies. Greenish below.

Distribution and habitats: Limited to high pH alkaline or saline areas with sparse vegetation of the Great Basin area, this species often occurs in

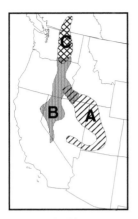

Map 43 Dark Saltflat Tiger Beetle, *Cicindela parowana*; **A**, *C. p. parowana*; **B**, *C. p. platti*; **C**, *C. p. wallisi*.

habitats that are neither wet nor closely associated with rivers or lakes.

Behavior: A solitary species, it often flies up out of white alkaline soil areas into low brush and sagebrush areas.

Seasonality: Adults are most common from April to June and again from July to September.

Larval biology: The larva has been described but its biology is unknown.

Festive Tiger Beetle, *Cicindela scutellaris* Say (Plate 11) [Map 44]

Description and similar species: Length 11–14 mm; robust, short-legged, and generally brightly colored, the Festive Tiger Beetle has elytra that are rounded at the rear end and relatively parallel-sided. This species shows the greatest geographical variation among populations of any tiger beetle species in North America with colors ranging from maroon and bright metallic orange to blue, green and black. However, males have a white labrum and females a dark or black labrum in all forms. Originally placed by Rivalier, together with the closely related Autumn Tiger Beetle, in its own taxonomic group, *Pachydela,* more recent molecular analysis indicates that these two species are indistinguishable from Rivalier's subgroup *Cicindela.* We follow Freitag by including the Festive Tiger Beetle in the genus *Cicindela.* The extremely similar Autumn Tiger Beetle was considered until recently to be a subspecies of the Festive Tiger Beetle, but detailed studies of seasonality, morphology of the middle tooth of the labrum, the segments of labial palpi and the brightness of the upper parts of the body justified its being separated into a different species (polymorphic with all black, dark blue or dark green forms). The male of the Autumn Tiger Beetle has a black labrum with two white spots that immediately separates it from the all-white labrum of the male Festive Tiger Beetle. Other similar species are detailed within the subspecies accounts below.

Subspecies and morphological variants: Seven distinctive populations are generally recognized as subspecies. In south Texas (Kenedy County) an apparently disjunct population of small, blue individuals may constitute an additional subspecies. At contact zones between most subspecies, considerable variation and intergradation are evident.

C. scutellaris scutellaris **Say**: With a smooth and brightly metallic surface on the upper parts, the dark green to blue head and thorax contrast vividly with the intense reddish-orange of the elytra. Some individuals show metallic green intruding onto the anterior parts of the elytra. Maculations are absent or occasionally present as small ivory-colored spots and a thin, short band along the outer edge of the elytra. This subspecies is the dominant form in the extreme western portion of the species' range, roughly west of the Missouri River. A broad contact zone with *lecontei* produces tremendous variability among individuals in eastern South Dakota, Nebraska and Kansas. The only species of tiger beetle likely to be confused with this form of *scutellaris* is the Splendid Tiger Beetle, which shares the green head and thorax and reddish elytra, but has a dull or matte surface on the elytra, not shiny or metallic.

C. scutellaris flavoviridis **Vaurie**: Usually lacking maculations or with two tiny ivory spots along the outer edge of the elytra, the highly metallic upper surface of this subspecies is diagnostic. Individuals are greenish-yellow approaching chartreuse in color. Some individuals of the Six-spotted Tiger Beetle approach this color, but the tapered rear end, rounded elytral edges, and presence of four to six small ivory spots on the elytra distinguish it from *flavoviridis*. This form of the Festive Tiger Beetle is restricted to central northern Texas and apparently does not intergrade with *rugata* to the east. It is very similar to the Green Claybank Tiger Beetle, with which it occurs in Texas.

C. scutellaris lecontei **Haldeman**: The upper surfaces of individuals in this population are duller and less shiny than the nominate form, and the reddish-maroon color (some individuals are an olive-green) is uniform from the head to the elytra. The broad ivory-colored border on the outer edge of the elytra can vary from complete to narrowly interrupted in back of the rear leg. This subspecies occurs throughout the northeastern and Midwestern part of the species' range. It is unlikely to be confused with any other tiger beetle species in North America. It intergrades broadly with nominate *scutellaris* and along narrower contact zones with three other subspecies. In the northern Midwest and prairie provinces, many individuals have the maculations coalesced into a distinct band along the entire edge of the elytra. Some think this character consistent enough to justify calling them a separate subspecies, *C. s. criddelei* **Casey**.

C. scutellaris rugata **Vaurie**: Confined to eastern Texas, northwestern Louisiana and southwestern Arkansas, this moderately shiny and metallic blue to blue-green form has no white maculations. It is separated to the east from *unicolor* by the Mississippi River flood plain but intergrades in the northwest of its range with nominate *scutellaris*, and in the northeast of its range with *lecontei*. Greenish forms somewhat resemble the Six-spotted Tiger

Beetle, but *rugata* lacks the six to four ivory spots on the elytra found in most Six-spotted Tiger Beetles, which also have a more tapered rear end. The Festive Tiger Beetle has a more noticeably domed profile on the elytra. Also, both sexes of the Six-spotted Tiger Beetle have a whitish labrum.

C. scutellaris rugifrons **Dejean:** This is the only polymorphic subspecies of the Festive Tiger Beetle, with black morphs more common in the north (eastern Massachusetts south into Eastern Shore of Maryland) and green more common to exclusive in the south. Also, coastal populations tend to be mostly green with fewer than 10% black. This proportion reverses itself inland where black often predominates. Individuals have a dull upper surface that is either black or green except for a bold white spot or triangle midway along the margin of each elytron, a small "C"-shaped maculation at the rear end of the elytra, and occasionally a small spot at the front part of the edge of the elytra. It occurs along the Atlantic seaboard south to the Carolinas, where it intergrades into the *unicolor* subspecies. A very narrow band of intergradation with *lecontei* runs through Connecticut and eastern Massachusetts.

C. scutellaris unicolor **Dejean:** Exhibiting a broad intergrade zone in North Carolina with the polymorphic *rugifrons*, the more distinct populations of *unicolor* in Georgia, Florida, and the Gulf Coast are all of the same color (monomorphic) with shiny blue, blue-green, or yellow-green upper parts and no white maculations. Only in northern Missouri and Tennessee do intergrades with *lecontei* become evident. The closely related species, the Autumn Tiger Beetle, occurs in the same habitat as *unicolor* from southern North Carolina to northern Florida, but primarily in the fall and not the spring when most *unicolor* are also active. In addition, the Autumn Tiger Beetle has a duller less metallic surface above, and it tends to have two common morphs, all dark-blue or all black, and more rarely a dark green form. Male *unicolor* have an all-white labrum but that of the male Autumn Tiger Beetle is black with two white spots.

C. scutellaris yampae **Rumpp:** Above metallic and smooth with the green head and thorax contrasting with the purplish-red elytra. Similar to the nominate subspecies, individuals of *yampae* are distinguished by a relatively broad and continuous ivory-colored band on the posterior half to three-fourths of the margin of each elytron. Isolated from other populations of *scutellaris, yampae* is found only in a small area of northwestern Colorado, where it occurs together with the *gibsoni* subspecies of the Big Sand Tiger Beetle and the *jordai* subspecies of Blowout Tiger Beetle.

Distribution and habitats: The species occurs over the entire eastern two-thirds of the United States and contiguous parts of southern Canada. It is, however, absent from the mountains of the Appalachians and from the lower Mississippi River flood plain and delta. It occupies sand dunes, dry

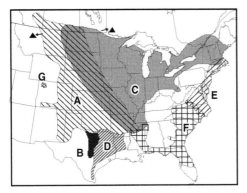

Map 44 Festive Tiger Beetle, *Cicindela scutellaris*; **A,**
C. s. scutellaris; **B,** *C. s. flavoviridis*; **C,** *C. s. lecontei*; **D,**
C. s. rugata; **E,** *C. s. rugifrons*; **F,** *C. s. unicolor*; **G,** *C. s. yampae.*

grassy sand blowouts, and sandy road cuts in the western parts of its range, and the sandy floor of open pine and oak–pine forests in the east and southeast. Over most of its range it can be found in these dry sandy habitats together with the Big Sand Tiger Beetle.

Behavior: Occupying habitats that are fully exposed to radiation from the sun and with no standing water for cooling, adults at mid-day frequently seek the shade of sparse vegetation or burrow into shallow holes in the sandy substrate until the afternoon temperatures fall. These burrows are also used to pass the night.

Seasonality: Adults are active in the spring from March to May and again in the fall from August to October. The life cycle is 2 years in most areas, and both adults and larvae overwinter.

Larval biology: Burrows occur in open patches of stabilized, deep sand or among patches of scattered, sparse vegetation. The burrows are relatively deep (25–80 cm). In many areas, larvae are heavily attacked by the bee-fly parasitoid, *Anthrax* (Fig. 7.2).

Autumn Tiger Beetle, *Cicindela nigrior* Schaupp
(Plate 11) [Map 45]

Description and similar species: Length 11–14 mm; above black, dark blue or dark green. Maculations are usually absent. Labrum black in females and black with two ivory white spots in males. Below the body segments are metallic blue-green to black. Extremely similar to and often occurring in the vicinity of the *unicolor* subspecies of the Festive Tiger Beetle, the two forms are separated by the duller coloration and larger size of the Autumn Tiger Beetle. Also the male Autumn Tiger Beetle has a black labrum with two white spots and the Festive Tiger Beetle male has an all-white labrum. Additional distinctions are possible only with microscopic examination of some physical differences. The Autumn Tiger Beetle has the middle point or tooth pointing down from the upper lip (labrum) shorter than the teeth on either side of it; the next to the end segment of the labial palpus is twice as thick as the end segment. Microhabitat and seasonal activities also serve to help distinguish the two species.

Subspecies and morphological variants: Black forms predominate in the southern part of its range and blue-green are more common in the northern part.

Distribution and habitats: Found in grassy areas of sand, this species is restricted to a narrow band along the eastern base of the southern Appalachian Mountains. While the Festive Tiger Beetle occurs primarily in loose sandy substrates in open areas of forest, the Autumn Tiger Beetle is found on tightly packed sandy areas with grass and rarely occurs into the forested areas.

Map 45 Autumn Tiger Beetle, *Cicindela nigrior.*

Behavior: The Autumn Tiger Beetle is more hesitant to fly than the Festive Tiger Beetle, but when it does fly, it usually goes into vegetation. The Festive Tiger Beetle usually flies again to open sand.

Seasonality: Unique among tiger beetles of the United States and Canada, and distinctively different from the spring–fall activity of the similar Festive Tiger Beetle, the Autumn Tiger Beetle is only active as adults in the fall, from September to November or rarely into early January

Larval biology: Larva unknown.

Oblique-lined Tiger Beetle, *Cicindela tranquebarica* Herbst (Plate 12) [Map 46]

Description and similar species: Length 11–15 mm; extremely variable in size and color above (reddish-brown, black, green) the two characters shared by most populations are the lack of a extended white line running along the outer edge of the elytra, and, if present, the rear of the front maculation is a long straight line angled away from the outer edge of the elytron, and with a rear tip that is not expanded. The width of the maculations ranges from thick to thin or even absent in individuals of some populations. Body segments below ranges in color from copper, metallic green, blue and purple to almost black.

Subspecies and morphological variants: Presently this wide-ranging species is divided into thirteen recognized subspecies, most of which are found in the mountainous west. Considerable variation, even within these geographic populations is compounded by complex and often broad zones of intergradation. Considerable disagreement surrounds the designation of many of these subspecies, and at least another fourteen subspecies names have been proposed at one time or another that we do not include here. Many of the subspecies we do include are tentative and may need to be reevaluated.

C. tranquebarica tranquebarica Herbst: Occurring throughout eastern North America, the nominate subspecies is the most widespread. The average length of individuals becomes gradually smaller from north to south. It is dark brown, gray or blackish above with thin to moderately wide maculations. Below it varies from metallic green and copper to purple. It occurs in an amazingly broad range of habitats that include, sandy beaches on oceans, lakes and rivers, mud flats, sand dunes, pine barrens, stubble fields, pastures, trails, paths, road side ditches, alkali flats, cement sidewalks, sand and gravel pits, golf courses, and almost any sparsely vegetated upland habitat. Intergrades broadly with *kirbyi* in the Midwest.

C. tranquebarica arida Davis: One of the most isolated and distinctive subspecies, this population is confined to the Death Valley area of California and Nevada and isolated from other populations of *tranquebarica*. No intergrades are known. Individuals are the smallest of the species with some males only 11 mm long. Above chartreuse green to olive-green (black in old individuals), with most individuals lacking maculations except for white tips to the elytra end. A few individuals exhibit thin maculations, and only then does the straight and angled front maculation makes it relation to other Oblique-lined Tiger Beetle populations obvious. Metallic green below. This subspecies is usually confined to green grass near water or temporary muddy and saline areas, and its escape flights are short and return quickly to the water's edge. Because of the intense heat of this area, its activity is almost completely limited to early morning in the early spring and more rarely in the late fall. Because of its isolation, unique habitat, behavior, body size and coloration, some experts consider this population a separate species. The subspecies *arida* shares a similar color pattern with *joaquinenesis*, but the latter occurs in the San Joaquin Valley of California, and its average size is larger.

C. tranquebarica borealis Harris: Above dark brown to black or dark green with thin maculations. Metallic green and copper to all purple below. Occupies wide range of habitats from water's edge to sage brush in the Northern Great Basin between the Rocky Mountains and Cascades. It confusingly intergrades with several other subspecies, and perhaps this entire population represents a zone of intergradation and not a distinct subspecies.

C. tranquebarica cibecuei Duncan: Range highly restricted to east central Arizona in the vicinity of the town of Cibecue in Navajo and Gila counties.

Black to navy-blue above with wide maculations. The rear end of the front maculation extends so that it almost touches the elbow of the middle maculation. The middle line is also attached at its base to a short but distinct line running along the edge of each elytron, but it does not reach the other maculations. Found only on sandy beaches of moderate to large mountain streams. Several authorities have combined this subspecies together with dark individuals from eastern California and Utah and called it *lassenica* Casey. Individuals from California and Utah, however, lack the uniquely extended maculations, and they are isolated from the Arizona population by subspecies *moapana* Casey and *diffracta* Casey. Based on these morphological and biogeographic inconsistencies, we retain the name *cibecuei* for the Arizona population and hypothesize that the dark individuals from California and Utah are more likely either older or melanistic individuals of other subspecies.

C. tranquebarica diffracta **Casey:** Above reddish-brown with wide maculations. Below the coppery thorax and metallic green abdomen contrast. Confined to sparsely vegetated beaches and flood plains along streams in southern Nevada, northern Arizona and New Mexico.

C. tranquebarica inyo **Fall:** Freshly emerged adults dark green and older individuals gradually turning darker above with wide maculations. The front maculation, however, is shortened to form a distinct gap between its rear end and the elbow of the middle maculation. The middle line, at its base, runs to the edge of the elytron but with little or no formation of a line extending forward and backward along the edge. Metallic green to dark purple below. Isolated to the Owens Valley of interior central California and adjacent Nevada. Old individuals in July often burnt brownish above and the maculations dirty, perhaps from extended exposure of their cuticular waxes to intense heat. Occurs on alkali mud flats, adjacent sandy areas, and near water.

C. tranquebarica joaquinensis **Knisley and Haines:** Green with reduced maculations, this recently described subspecies is similar to *arida* but larger and restricted to the San Joaquin Valley of California. It intergrades with *vibex* on the western slopes of the valley and occurs on only a few remnant alkali flats.

C. tranquebarica kirbyi **LeConte:** Above brown with wide maculations. Below metallic, copper, green, or purple. Occurs west of the Rocky Mountains in the Great Plains of Canada and central United States. Intergrades widely with the nominate form in the Midwest. Occupies wide range of moist and dry habitats.

C. tranquebarica moapana **Casey:** Known only from central eastern Nevada and adjacent Utah, this subspecies has blackish upperparts with wide maculations. Below distinctly bicolored with the front half bright copper and the rear half metallic bluish- green. Occurs primarily on sandy beaches of large creeks and small mountain rivers.

C. tranquebarica parallelonata **Casey:** Above dark green, brown or blackish, older individuals darker, maculations wide. Some differences in upperpart color may be due to darkening with age and exposure to sunlight and abrasion from burrowing into sandy substrates. Metallic green to greenish-blue below. Found only in southern Nevada and adjacent Utah and eastern California, this subspecies occupies edges of alkaline pools, mud flats bordering lakes, river banks, open muddy ground often covered with stones of basalt that match the size and color of these tiger beetles.

C. tranquebarica sierra **Leng:** Above dark green with such thin maculations that the front one is often absent. Metallic green-blue below. Stream banks and moist soil with grass at high elevation (2000 m) in the Sierra Nevada Mountains of eastern California, where it is most likely to be found in June and July.

C. tranquebarica vibex **W. Horn:** Above brown, green or dull coppery-green with thin maculations in which the first one is often reduced to a short line. Metallic purple to green below. Occurs from western British Columbia south to California. Generally specimens west of the Cascades are green and those east of the Cascades are brown to dark green, but mixed populations are found throughout the area. Those east of the Cascades may be intergrading with *borealis*. This subspecies is much rarer west of the Cascades and occurs in habitats such as sea beaches, mud flats along lakes, ponds, streams and irrigated pastures. East of the Cascades it is more common and is found on salt flats, open dry ground, roads and forest openings. Because there is so much variation in color patterns, we can find no justification to separate the generally green *roguensis* E. D. Harris from the coppery-green *vibex*.

C. tranquebarica viridissima **Fall:** Above light green to blue-green with thin to moderate maculations. Below metallic green to blue-green. This population is restricted to coastal southern California, primarily in Orange and western San Bernardino counties, and most specimens were taken from orange groves near Anaheim in the early 1970s and along the Santa Ana River (Riverside County) in the 1980s. Here it was active from January to February and again in October and November. The subsequent urbanization and channelization of appropriate habitat may have extirpated this population.

Distribution and habitats: Generally the eastern and northern populations occupy a very wide range of habitats, but it is strangely absent in Florida from

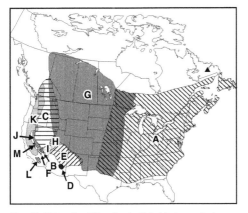

Map 46 Oblique-lined Tiger Beetle, *Cicindela tranquebarica*; **A**, *C. t. tranquebarica*; **B**, *C. t. arida*; **C**, *C. t. borealis*; **D**, *C. t. cibcuei*; **E**, *C. t. diffracta*; **F**, *C. t. inyo*; **G**, *C. t. kirbyi*; **H**, *C. t. moapana*; **I**, *C. t. parallelonata*; **J**, *C. t. sierra*; **K**, *C. t. vibex*; **L**, *C. t. viridissima*; **M**, *C. t. joaquinensis*.

all but the extreme northwestern part of the panhandle. The desert southwest populations are more restricted to mud flats, moist alkali areas and beaches of streams and lakes.

Behavior: Depending on the population, this species can be solitary to gregarious. It is a strong flier, and individuals are occasionally attracted to lights at night. Its escape flight often involves wobbles from side to side before landing. Both adults and larvae overwinter in most areas.

Seasonality: Specimens have been found every month of the year except December. However, climate, elevation and latitude influence seasonality locally. Adults are most active between April and May and again between September and October. The life cycle in most areas is 2 years.

Larval biology: Burrows are along flood plains or in moist soil in the west and in dry, bare clay areas in the east. Often the burrows are present in high densities and are heavily attacked by the parasitoid bee-fly, *Anthrax* (Fig. 7.2). Burrow depths are shallow to moderate (22–50 cm) depending on soil composition.

Appalachian Tiger Beetle, *Cicindela ancocisconensis* T. W. Harris (Plate 13) [Map 47]

Description and similar species: Length 14–16 mm; above brownish-bronze with green reflections on head; three thin maculations on each elytron, the middle line has a long base along the edge of the elytra, and it is interrupted at the "elbow" in some individuals; the front maculation is shortened at its rear end and does not curl forward. The elytra are long and parallel-sided; below metallic dark green to purplish-blue. Also diagnostic in the hand are three teeth on the labrum. Often found together with the similar Bronzed Tiger Beetle, which has a "C"-shaped front maculation, more-rounded elytral sides, and a single tooth on the labrum. Another similar species, the Oblique-lined Tiger Beetle, is darker brown above and has a longer, straighter, and angled front maculation. It also has almost no baseline running along the outer edge of the elytra connecting to the middle line.

Subspecies and morphological variants: No geographically distinct populations are known, but populations in northern parts of its range have more reduced maculations.

Map 47 Appalachian Tiger Beetle, *Cicindela ancocisconensis.*

Distribution and habitats: The Appalachian Tiger Beetle is confined primarily to low mountain rivers in hilly areas in the eastern United States and southeastern Canada where it occupies sand bars, shaded beaches, and gravel areas along forested river courses. Apparently rare or extirpated from most of its former range along the Ohio River. Old records indicate that at one time it may have occurred as far west as Indiana and perhaps Illinois.

Behavior: Occurring in small localized colonies, it usually is found within 60 m of the water's edge. Rarely, however, it is found in wet sandy areas up to 2.5 km from the nearest water. Although it occurs regularly with the Bronzed Tiger Beetle and the Twelve-spotted Tiger Beetle, the Appalachian Tiger Beetle is usually less common and more likely to be present in beaches with heavier vegetative cover. With a naturally patchy distribution throughout its range, its absence from many historical sites makes it difficult to determine how much impact dam building and water pollution may have had on these changes from its former range. Alternatively it could be adapted to river flooding and habitat scouring to abandon these areas and move quickly to newly formed habitats some distance away.

Seasonality: Adults are active from April to June and again from mid-July to September, but fall activity in most areas is much reduced or even absent.

Larval biology: Burrows in sandy-loam soil of upper flood plains, often far away from water's edge. Burrow depth 8–20 cm.

Blowout Tiger Beetle, *Cicindela lengi* W. Horn
(Plate 13) [Map 48]

Description and similar species: Length 12–15 mm; above red to reddish-brown (green, blue, or dark maroon in Canada), maculations wide and often connected along the edge of the elytra; front maculation angles

119

away from the elytral edge in a straight line with no hook on its rear end. Below metallic green with dense white hairs. Most similar to the Big Sand Tiger Beetle, with which it shares the same sandy habitats in the west. Blowout Tiger Beetle has a narrower body, longer and narrower labrum, and a longer front maculation. The Oblique-lined Tiger Beetle is darker above with a thinner front maculation. The Crimson Saltflat Tiger Beetle is smaller, much shinier above, and does not occur in dry sandy areas.

Subspecies and morphological variants: Three subspecies are recognized, and they are distinguished by the extent of the white maculations covering the elytra and the color of the underside of the thorax. Considerable intergradation of characters makes the placement of many specimens into a subspecies difficult.

C. lengi lengi W. Horn: The underside of the thorax and abdomen are the same color, metallic blue-green. The three maculations of the elytra are wide but distinct. If they are joined together, it is only along the edge of the elytra. This form is found more consistently in the southeastern part of the species' range.

C. lengi jordai Rotger: Occurs in the Four Corners area of the southwestern United States. Above uniformly bright red and the maculations are broadly coalesced to form a wide white band covering two-thirds of each elytron.

C. lengi versuta Casey: Typical individuals of this subspecies are similar to the nominate form but have a distinctly coppery underside of the thorax that contrasts with the metallic green underside of the abdomen. This form is found more consistently in the northern part of the species' range. Individuals that have normal maculations but are all blue, green or even black above are regular in the Prairie Provinces. In Manitoba, most individuals have the reddish head and thorax replaced with a dull greenish color.

Distribution and habitats: Confined to dry sand and upland sand-clay areas with sparse vegetation. In the southwest it is associated with sandy banks of dry washes in canyon beds and in other parts of its range it is found in sand dune margins, blowouts, barren flats, roadsides, and occasionally sand bars along

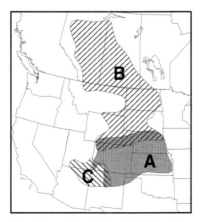

Map 48 Blowout Tiger Beetle, *Cicindela lengi*; **A,** *C. l. lengi*; **B,** *C. l. versuta*; **C,** *C. l. jordai*.

rivers. In the far north of its range it occurs on sandy roads through open coniferous forests.

Behavior: This species, while dependent on sandy areas, apparently can survive in much smaller areas of sand than most of the other sand-inhabiting species. In areas undergoing revegetation, the Blowout Tiger Beetle is often the last tiger beetle to disappear from the area. It regularly occurs together with the Big Sand Tiger Beetle and the Festive Tiger Beetle.

Seasonality: Adults are active from late April to June and from August to September.

Larval biology: Burrows are found in low densities in open, dry sandy soils on both slopes and flat ground, and they are quite deep (70–130 cm).

Short-legged Tiger Beetle, *Cicindela tenuicincta* Schaupp
(Plate 13) [Map 49]

Description and similar species: Length 12–16 mm; above dark brown to black with some olive-green reflections; three maculations wide and distinct with first two connected by a line along outer edge of elytron; rear of front maculation straight and angled away from outer edge of elytron, ends in an expanded wedge or arrowhead shape; below metallic greenish-black. Noticeably short rear legs. Occurs together with the similar Oblique-lined Tiger Beetle, which, however, lacks or has a reduced connecting line along the edge of the elytra as well as no expended rear tip to the front maculation.

Map 49 Short-legged Tiger Beetle, *Cicindela tenuicincta.*

Subspecies and morphological variants: There are no known distinct geographical races over the limited Great Basin range of this species.

Distribution and habitats: Patchy distribution limited to moist mudflats, often alkali with no vegetation, beaches along lakes, and ponds and streams of the Great Basin.

Behavior: Because of its excellent camouflage in the often expansive and continuous habitat, this species is difficult to see until it flies, and then it often goes long distances. It is occasionally attracted to lights at night.

Seasonality: Adults are active in the early spring from May to June and then again from mid-July to September. In northern California, however, it is a summer species and active as adults from June to September.

Larval biology: Larva unknown.

Western Beach Tiger Beetle, *Cicindela latesignata* LeConte
(Plate 13) [Map 50]

Description and similar species: Length 12–13 mm; above black to blackish-green with very wide but distinct maculations that have connections that extend and often coalesce along the elytral edge. On some individuals these maculations are so expanded that they coalesce considerably along their entire lengths. Body segments below dark metallic green to blue-green. Frequently found in the same coastal habitat, the similar Senile Tiger Beetle has thinner maculations with the first one reduced to a short and shallow "C" shape, and the middle one with no extension along the outer elytral edge. These two species are usually active at different times of the year.

Subspecies and morphological variants: Only the nominate subspecies occurs in the United States. An additional subspecies that is reddish above, is found in northwestern Mexico (Baja California and Sonora), *C. l. parkeri* Cazier.

Distribution and habitats: Restricted to coastal sea beaches, bays, estuaries, salt marshes and alkali sloughs in southern California from Los Angeles County south into Mexico. Early records from Santa Barbara County are likely erroneous. Destruction of its narrow habitat in the United States has eliminated most of its former range except for protected areas and military installations.

Behavior: Often occurs together in large numbers pursuing concentrations of prey on tidal flats, especially shore flies and small crustaceans.

Seasonality: Adults are active from May to October but most common in July.

Larval biology: Tidal flats just above high tide in mud or sand. Burrows usually curved to become nearly horizontal at the bottom.

Map 50 Western Beach Tiger Beetle, *Cicindela latesignata.*

Williston's Tiger Beetle, *Cicindela willistoni* LeConte
(Plate 14) [Map 51]

Description and similar species: Length 10–13 mm; highly variable color above from shiny reddish or brown to dark green or coppery. Maculations vary from absent in some populations to broadly coalesced to cover most of the elytra in other populations. Populations east of the Rocky Mountains tend to have parts of their heads thickly covered with white, hair-like setae and those west of the mountains have few hairs on the head.

Subspecies and morphological variants: The patchy distribution of its specialized habitat, moist saline soils, has evidently led to long isolation of several of these populations. Eight subspecies are recognized, and zones of intergradation between some populations are absent or often narrow.

C. willistoni willistoni **LeConte:** Above reddish-brown with maculations broadly coalesced, especially the front two. Below metallic dark green-purple. Found in southeastern Wyoming and perhaps adjacent Colorado. It intergrades to the west with *C. w. echo* Casey.

C. willistoni echo **Casey:** Above dark brown with wide maculations that are not connected along the elytral edge. Below metallic green-blue or rarely coppery. Found in the Great Basin from Wyoming and southern Oregon to California and Utah. Individuals in the northern Mojave Desert of California tend to be small. They may represent intergrades with *C. w. praedicta* or perhaps even an unrecognized subspecies.

C. willistoni estancia **Rumpp:** Above coppery-brown with whitish maculations completely coalesced to cover all but a "U"-shaped area at the front of the elytra and a dark line down the length of the middle. A few individuals have the coalesced white maculations less complete, but even this is greater than in any other subspecies. Below dark greenish-blue. Restricted to several large saline lakes in Torrance County of central New Mexico.

C. willistoni funaroi **Rotger:** Above reddish with wide maculations that are connected along the edge of the elytra. Below dark bluish-green. Found only in Sandoval County, in northwestern New Mexico, individuals tend to be smaller than in other populations (10–11 mm).

C. willistoni hirtifrons **Willis:** Above bronze or brown with wide but separate maculations with usually only first two maculations connected at outer elytral edge. Below metallic blue-green to coppery. Occurs patchily in saline areas across the southern Great Plains.

C. willistoni praedicta **Rumpp** Above dark, oily green to greenish-blue with no maculations except for small white tips at end of elytra. A few individuals exhibit a thin middle maculation that is typical of maculated subspecies of

Williston's Tiger Beetle. Below metallic dark purple-blue. Found only in two adjacent valleys (Shoshone and Tecopa) in the Death Valley area of east central California and adjacent Nevada. Extremely similar to and co-occurring with the nominate form of the Great Basin Tiger Beetle, which has larger white tips at the end of the elytra, the elytra themselves are flatter in profile, and greener to bluish-green above.

C. willistoni pseudosenilis W. Horn: Above dark green to greenish-blue with wide maculations that are not connected along the elytral edge in most individuals. Below dark metallic bluish-green. Confined to the Owens and adjacent Panamint Valley of east central California. In the southern part of these valleys there is some intergradation with *C. w. echo*.

C. willistoni sulfontis Rumpp: Above dark green or dark reddish-brown with wide maculations broadly connected along the elytral edge. Body segments below metallic purple-blue. Known only from the Sulphur Springs Valley, Cochise County, of southeastern Arizona.

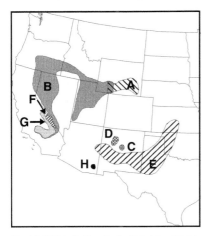

Map 51 Williston' s Tiger Beetle, *Cicindela willistoni*; **A**, *C. w. willistoni*; **B**, *C. w. echo*; **C**, *C. w. estancia*; **D**, *C. w. funaroi*; **E**, *C. w. hirtifrons*; **F**, *C. w. praedicta*; **G**, *C. w. pseudosenilis*; **H**, *C. w. sulfontis*.

Distribution and habitats: Limited to salt and alkali flats including muddy areas near temporary ponds, reservoirs, streams and seepages. Usually on wet mud but also in drier areas among widely spaced grass clumps.

Behavior: During the heat of the day, it seeks out shady depressions in the mud. Often flies short distances spontaneously, evidently to find cooler areas, but its flights to escape danger can be long.

Seasonality: Apparently adults of some populations are spring–fall active (*pseudosenilis* March–May, July–November; *hirtifrons* April–June, September–October; *funaroi* April–May, September) and others are summer active (*estancia* May–September; *praedicta* March–May; *sulfontis* June–August; *willistoni* May–July). This may be an adaptation to differing summer rainy seasons. Most subspecies have a 2-year life cycle.

Larval biology: Burrows are generally located in clay or sandy-clay soils on or along the edges of open salt flats where there is permanent moisture and poor drainage. Burrow depths are 14–22 cm. Larvae of at least three subspecies, *echo, hirtifrons*, and *sulfontis*, build peculiar chimney-like turrets (Fig. 7.1) over the larval burrows to attract prey and thermoregulate. These turrets are obvious standing several centimeters high in an otherwise flat lake bed or salty mud flat.

Senile Tiger Beetle, *Cicindela senilis* G. H. Horn
(Plate 14) [Map 52]

Description and similar species: Length 10–12 mm; above dark brown to blackish, three thin maculations with no line connecting them along the outer elytral edge. The front maculation is reduced to a small and shallowly curved "C" shape; the middle maculation reaches the elytral edge but has no base running along the edge of the elytra. Below metallic green to blue-green. Although not a unique character, the Senile Tiger Beetle gets its name from the abundant white hairs covering its head. The similar Western Beach Tiger Beetle occurs in the same coastal habitats but has wider maculations, especially the first one, which has its rear portion extended in a straight line and angled away from the outer edge of the elytra. The middle maculation has a distinct base running along the outer edge of the elytra that coalesces with the front maculation in some individuals. These two species are usually active at different times of the year.

Subspecies and morphological variants: The populations from Los Angeles, Orange and San Diego counties, California, have been separated by some experts from the nominate species on the basis of greener upperparts into the subspecies ***C. senilis frosti*** **Varas-Arangua**.

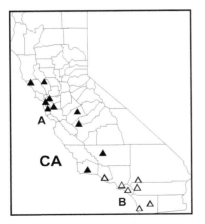

Map 52 Senile Tiger Beetle, *Cicindela senilis*; **A,** *C. s. senilis*; **B,** *C. s. frosti*.

Distribution and habitats: Endemic to California and adjacent northern Mexico. Formerly this species occurred in coastal salt marshes and tidal mud flats as well as interior alkali mud flats from San Diego County north to Sonoma and Lake counties. Urbanization has apparently severely reduced or extirpated all but a few of the coastal populations in San Luis Obispo and Sonoma counties. In the interior of Southern California, this species is more common, but even here these populations are threatened by continuing urban development of the habitat.

Behavior: Adults overwinter in shallow underground galleries, usually under flat rocks at the edge of salt marshes.

Seasonality: Adults are active from February to June and then again from August to October.

Larval biology: Larva unknown.

Great Basin Tiger Beetle, *Cicindela amargosae* Dahl
(Plate 15) [Map 53]

Description and similar species: Length 10–12 mm; above dark green, blue-green, or black with no maculations except for small white tips at the end of the elytra. The upper surface of greenish individuals is iridescent and shiny and that of black individuals is dull or matte. Below metallic blue or black with some populations showing coppery reflections. The green forms of eastern California and adjacent Nevada occur side by side with the similar *praedicta* form of the Williston's Tiger Beetle and can be distinguished only with careful inspection. The Great Basin Tiger Beetle is metallic green above with elytra flatter in profile. The white spots on the tip of the elytra are larger and more distinctive. The *praedicta* form of Williston's Tiger Beetle is oily green in appearance above with elytra distinctly domed in profile. Black forms of the Great Basin Tiger Beetle are similar to black forms of the Cow Path Tiger Beetle, which has middle and front maculations present. The Cow Path Tiger Beetle also rarely occurs near water courses or on salt flats. Black forms of the Long-lipped Tiger Beetle can be separated by their maculations, extremely long labrum and their absence from water edges and salt flats.

Subspecies and morphological variants: At one time the Great Basin Tiger Beetle was considered a subspecies of the Williston's Tiger Beetle and then later a subspecies of the Senile Tiger Beetle. Recent comparative studies of their morphology indicate it is best considered a closely relaed but separate species. Two subspecies of the Great Basin Tiger Beetle have been described. Some populations in northern Nevada and California have dark green upper surfaces with a dull or matte texture. Other populations in southern Nevada have various percentages of individuals that are dull black and others that are shiny green. Black individuals are usually smaller than green ones.

C. amargosae amargosae **Dahl:** Above dark blue-green to green. This subspecies is restricted to the Death Valley area of eastern California and intergrades with *nyensis* in the adjacent Amargosa River valley to the east. Here a mixture of dark green, dark blue and black individuals occurs.

C. amargosae nyensis Rumpp: Above black. This subspecies occurs in a narrow band of scattered saline habitats along the western edge of the Great Basin. It intergrades with the nominate subspecies in southwestern Nevada and adjacent California along the Amargosa River Valley.

Distribution and habitats: The highly fragmented populations are restricted to grassy areas in saline habitats along edges of desert rivers, mud flats and shores of temporary ponds and lakes, salt flats and moist areas with isolated grass clumps.

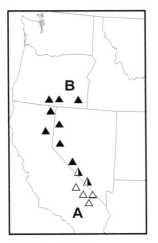

Map 53 Great Basin Tiger Beetle, *Cicindela amargosae*; **A**, *C. a. amargosae*; **B**, *C. a. nyensis*.

Behavior: Adults often concentrate in the limited moist soil areas of the deserts in which they live. Bright green individuals have escape flights that are straight and long (2 to 6 m). They land on moist, open soil and immediately begin to run. Dull black individuals have short escape flights (<2 m) that rise sharply up and then end abruptly, usually in a grassy patch. They usually do not move for some time after landing.

Seasonality: A summer active species, northern populations are active from April to October. Southern populations are restricted to April and May when snow melt from the nearby mountains fills intermittent or ephemeral desert rivers.

Larval biology: Larva unknown.

American Tiger Beetles, Genus *Cicindelidia*

Confined to the Western Hemisphere, this genus contains more than 80 species, most of which occur in Mexico and Central America. Nineteen of these species occur in the United States and Canada, and an additional species is hypothetical for its occurrence here. The characters of the genus include eyes less protruding and legs shorter than most other groups within the tribe Cicindelini. Many of the species expose contrasting red-brown to bright orange abdomens when their elytra are spread in flight. Most species of *Cicindelidia* are active as adults only during the summer.

Black Sky Tiger Beetle, *Cicindelidia nigrocoerulea* LeConte (Plate 15) [Map 54]

Description and similar species: Length 9–12 mm; above dull olive-green, dark blue, or black. Many populations include individuals of several colors. Maculations vary within populations from completely absent to broken dots to a distinct white band running the length of the outer edge of the elytra. Below dark metallic purple to black. The green forms with dots or no maculations can be confused with western green forms of the Punctured Tiger Beetle, which is shinier above and has a thinner more slender body shape, especially in the thorax. Also, the rear tips of the elytra come to a point in the Punctured Tiger Beetle but are rounded in the Black Sky Tiger Beetle. Below, the white hair-like setae of the Punctured Tiger Beetle are much more extensive, and, in the hand, the Black Sky Tiger Beetle has three distinct teeth on the labrum edge while the Punctured Tiger Beetle has but one in the center. Along the Mexican border in Arizona, New Mexico and west Texas, the similarly polymorphic Horn's Tiger Beetle also occurs in similar grassland habitats. It, however, is much shinier above, has a more domed elytral profile and a black labrum. The southern Texas population of the Black Sky Tiger Beetle is confusingly similar to the Cazier's Tiger Beetle, but they occur in different habitats and the bright orange abdomen of the Cazier's Tiger Beetle can be seen from underneath or from above when the elytra are spread to expose it in flight.

Subspecies and morphological variants: Three subspecies are recognized, and they are distinguished primarily by the extent of maculations on the elytra and to some degree the color above.

C. nigrocoerulea nigrocoerulea LeConte: Above dull olive-green, dark blue or black with no maculations or a few inconspicuous white dots on the elytra. Dark blue morphs tend to predominate in most populations. Occurs from central Mexico north into the southwestern United States.

C. nigrocoerulea bowditchi Leng: Above uniformly olive-green with highly variable maculations ranging from a white band around outer edge of elytra and/or diffusion of white spots or complete but thin maculations on elytra. This subspecies occurs in the northern part of the species range in Colorado and adjacent New Mexico. There is a narrow zone of intergradation with the nominate subspecies in north central New Mexico.

C. nigrocoerulea subtropica Vogt: Above black with maculations reduced to small spots. Black below. This subspecies is restricted to Hidalgo and Cameron counties in the lower Rio Grande Valley of south Texas. It occurs in bare patches of soil among short grass, bare soil of citrus groves, edges of plowed fields, and along drainage ditches. Active in the late fall and early winter, adults emerge for only a short period following measurable rainfall. It has also been reported in adjacent Mexico.

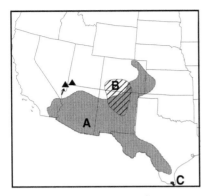

Map 54 Black Sky Tiger Beetle, *Cicindelidia nigrocoerulea*; **A**, *C. n. nigrocoerulea*; **B**, *C. n. bowditchi*; **C**, *C. n. subtropica*.

Distribution and habitats: Occurs in a wide variety of habitats but most often on moist upland soils, such as in grasslands, bare soil areas, mud flats, and along temporary pond edges in the southwest.

Behavior: To escape danger frequently flies from bare soil areas to hide in dense grass. During very hot weather, this species may spend most of the day hiding in dense vegetation along the edge of a body of water. Often concentrates around muddy water edges but during rainy periods moves far into grasslands away from standing water and becomes more solitary. Found together most frequently with the Punctured Tiger Beetle, Horn's Tiger Beetle, and Large Grassland Tiger Beetle in upland areas and with Western Red-bellied Tiger Beetle, Ocellated Tiger Beetle and Wetsalts Tiger Beetle along the water's edge.

Seasonality: Primarily a summer-active species (June–September), but in south Texas it is a fall-active species (September–November).

Larval biology: Burrows in sandy-clay soils in grasslands, often at the base of a clump of grass.

Horn's Tiger Beetle, *Cicindelidia hornii* Schaupp (Plate 15) [Map 55]

Description and similar species: Length 11–15 mm; above very shiny and smooth black, dark blue or dark green. No maculations. Most populations include all three color forms, but black is usually the most common form. Elytra extremely domed in profile. Black to dark purple below. Labrum black in both males and females. Overall appearance similar to nominate forms of the Black Sky Tiger Beetle, but the upper surfaces of the Black Sky Tiger Beetle are dull, the elytra are flatter in profile, and the labrum is white.

Subspecies and morphological variants: Within the United States only the nominate subspecies occurs. One other subspecies, *C. h. scotina* Bates, is all black and confined to the highlands of central Mexico.

Distribution and habitats: An upland species, it occurs on open, dry grasslands or hillsides with clay or loamy soil sparsely covered with grasses. Most easily seen when it emerges into bare patches of soil.

Map 55 Horn's Tiger Beetle, *Cicindelidia hornii*.

Behavior: Normally a weak flier, this species runs from grass clump to grass clump or makes short flights to land in dense grass clumps. If pursued persistently, however, an individual will suddenly fly up 5 to 10 m in the air and let the wind take it off across the grasslands sometimes for 500 m or more. Often hide under cow droppings as soil dries out. Interruptions in rainfall and desiccation of soils for more than a week cause the adults to become inactive. Most often found together with Large Grassland Tiger Beetle, Black Sky Tiger Beetle, White-striped Tiger Beetle, and Grass-runner Tiger Beetle.

Seasonality: Adults have been observed from May to November, but most are active during the summer rainy season June to September. Only the larvae overwinter, and the life cycle is 2–3 years.

Larval biology: The larval burrows are usually at the base of a grass clump or in a bare patch of soil that forms a depression temporarily to trap rain water.

Large Grassland Tiger Beetle, *Cicindelidia obsoleta* Say
(Plate 16) [Map 56]

Description and similar species: Length 15–20 mm; above dull black, brown, green and rarely dark blue with maculations that are absent in some populations, reduced to incomplete lines and dots in others, and thin but complete in others. If maculations are present, usually only the last one reaches the edge of the elytra, however individuals in some populations can have maculations partially connected by a white line along the outer elytral edge. Below metallic purple to dark green. Because of its large size, not easily confused with any other grassland species.

Subspecies and morphological variants: Presently divided into six subspecies, of which two (*C. o. juvenilis* W. **Horn** and *C. o. latemaculata* **Becker**) are restricted to Mexico. The subspecies are distinguished primarily on the basis of color pattern above and in some cases by average size.

C. obsoleta obsoleta Say: Above usually black or occasionally green but with no maculations. This subspecies occurs from Mexico north into the western part of the Great Plains. Broad zone of intergradation with *santaclarae* in central New Mexico, west Texas and Colorado and with *vulturina* in east Texas.

C. obsoleta vulturina LeConte: Above black to olive-green with thin maculations often reduced to spots and short lines. On some individuals a thin line along the edge of the elytra connects the first two maculations. The middle line, if present, is bent back at a sharper angle than in other subspecies. Coastal and western populations tend to have greatly reduced or no maculations, and northern populations tend to have complete or nearly complete maculations. An isolated population in southern Missouri and central northern Arkansas has complete or near complete maculations and virtually all individuals are dark green above.

C. obsoleta santaclarae Bates: Above green, brown or black with maculations broken into dots and commas. In many populations all three color forms are present together. This subspecies occurs in northwestern Mexico north to Arizona and western New Mexico and southwestern Colorado. Broadly intergrades with nominate populations in central New Mexico, west Texas and Colorado.

C. obsoleta neojuvenilis Vogt: The smallest subspecies (13–16 mm) of the Large Grassland Tiger Beetle, black above with green margins. Front maculation reduced to two dots; middle maculation a wavy line angling away from and not touching the outer edge of elytra. The rear maculation a line along the edge at the tip of the elytra connected to a single dot. An isolated population found only in the lower Rio Grande Valley of South Texas (Maverick County south to Hidalgo County), adults are active following rains in the late fall and early winter. Few specimens have been collected or observed since the original series was collected in 1946.

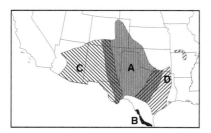

Map 56 Large Grassland Tiger Beetle, *Cicindelidia obsoleta*; **A**, *C. o. obsoleta*; **B**, *C. o. neojuvenilis*; **C**, *C. o. santaclarae*; **D**, *C. o. vulturina*.

Distribution and habitats: An upland species that is rarely found around standing water, The Large Grassland Tiger Beetle occurs in grasslands, pastures, meadows, grassy alluvial slopes and hillsides with bare patches of soil.

Behavior: Extremely wary, the Large Grassland Tiger Beetle flies up from danger quickly and is difficult to approach. It is so large that the buzzing wings often can be heard taking off in flight.

They fly long distances and often land in grassy clumps where they will then quickly run from their landing site and hide at the base of a grass clump. If they land in grass too thick, their large size makes it difficult for them to maneuver and run quickly. Individuals often hide under cow droppings as the soil dries out.

Seasonality: Primarily a summer species, the precise timing of summer rains is important for determining activity periods. Over its entire range active as adults from June to October, but in Colorado, New Mexico and Arizona mainly July to August. In south Texas it is usually active only in October to January. Life cycle 2–3 years. The isolated population in Missouri and Arkansas is active in September and October.

Larval biology: Burrows are large and obvious but shallow (6–10 cm). Found in gravelly to loamy soil in bare ground among grass tufts in well drained areas, usually on slopes. Tunnel often spirals around rocks and other objects.

Punctured Tiger Beetle, *Cicindelidia punctulata* Olivier
(Plate 15) [Map 57]

Description and similar species: Length 11–13 mm; a highly variable species, it can be black, brown, olive or metallic green, blue-green, or rarely bright blue above with maculations largely absent, reduced to small spots and short lines, or rarely complete. Tip of closed elytra comes to a broad point. Body narrow, especially the thorax. Below metallic green, blue, and copper. In the hand the labrum has only a single middle tooth. Two rows of shallow but distinct pits run parallel along either side of the inner elytral edges. In the west most easily confused with the Black Sky Tiger Beetle, which has a rounded end of the closed elytra, a broad body and thorax, and in the hand three teeth on the labrum. In the east the Punctured Tiger Beetle can be confused with the nominate subspecies of the Eastern Red-bellied Tiger Beetle, which has a contrasting orange abdomen.

Subspecies and morphological variants: Three subspecies are recognized, but one of these, *C. p. catharinae* **Chevrolat**, is restricted to the central highlands of Mexico. These subspecies are separated primarily on the basis of the color of their upper surfaces. An interesting population of mostly brown and well maculated individuals occurs in southeastern Wyoming and northwestern Colorado.

C. punctulata punctulata Olivier: Above black to dark olive, maculations reduced to a few small spots and a distinct short but straight line extending inward on the outer tip of the elytra. The nominate subspecies ranges from the Atlantic Coast to the Rocky Mountains, where it intergrades over a larger area with the western forms.

C. punctulata chihuahuae Bates: Above bright metallic green to green-blue with maculations absent or reduced to small spots. This subspecies occurs in the southwestern United States and northwestern Mexico. Occasionally some areas such as the San Rafael grasslands of Santa Cruz County, Arizona, have entire populations that are black. In parts of southwestern Colorado and eastern Utah, many individuals are bright blue above. This form is considered a separate species by some experts.

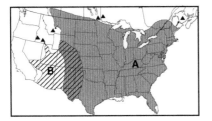

Map 57 Punctured Tiger Beetle, *Cicindelidia punctulata*; **A**, *C. p. punctulata*; **B**, *C. p. chihuahuae.*

Distribution and habitats: One of the most widely ranging species in North America, the Punctured Tiger Beetle in the east is found mainly in upland habitats with dry, hard-packed soils and sparse grasses present. These habitats include dusty roads, old trails, pastures, roadside ditches, strip mines, eroded gullies, city lawns, gardens, crop fields, sidewalks, parking lots, rock hillsides, sand pits, and dunes. In the west it occurs most regularly in the vicinity of water, such as the edge of rivers, lakes, marshes, irrigation ditches, temporary ponds, alkali mud flats from lowland prairies to mountain tops.

Behavior: A solitary species in dry upland habitats of the east, it tends to be more gregarious gathering around limited wet areas in the west. It is wary but a weak flier with short escape flights. Commonly attracted to lights at night. Because it so commonly occurs in agricultural areas, there is evidence that some populations may have developed resistance to pesticides.

Seasonality: A summer active species, adults can be found from April to November, but most are active July to August. In the Southwest, summer rains limit activity from July to September. Overwinters as larvae and has a 1- or 2-year life cycle.

Larval biology: Burrows are usually in hard-packed sand, clay, or loam with sparse grass. Burrow depths range from 14 to 40 cm.

Thin-lined Tiger Beetle, *Cicindelidia tenuisignata* LeConte
(Plate 17) [Map 58]

Description and similar species: Length 11–12 mm; above brown with green edges. Maculations thin but complete. The long middle maculation curves distinctively rearward with no sharp "elbow". It is attached to a relatively long line running along the edge of the elytra. The distinctive shape of the maculations makes it unlikely to confuse this species with any other in the United States and Canada. Greenish-copper below.

Subspecies and morphological variants: No geographical variants are known for this species.

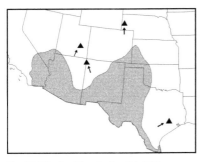

Map 58 Thin-lined Tiger Beetle, *Cicindelidia tenuisignata*.

Distribution and habitats: Occurs near water on the muddy or sandy edges of rivers, lakes, temporary ponds, and estuaries across much of the southwest.

Behavior: Usually solitary or in small numbers among other more abundant tiger beetle species such as the Wetsalts Tiger Beetle, Ocellated Tiger Beetle, and Rio Grande Tiger Beetle. However, at times it can be present in large numbers. This species apparently has broad powers of dispersal as errant individuals have been found as far north as western Nebraska. However, it has apparently not become established in these more northern areas. In desert areas of the southwest, it uses this dispersal ability to quickly colonize wet areas following a break in a drought. It as quickly disappears from some areas for several dry years. It is commonly attracted to lights at night.

Seasonality: Adults have been found from May to November but are most common June to September over most of its range. Overwinters as larvae and has a 2-year life cycle.

Larval biology: Burrows are in moist sandy or muddy soil near the water's edge.

Red-lined Tiger Beetle, *Cicindelidia fera* Chevrolat
(Plate 17) [Map 59]

Description and similar species: Length 10–11 mm; above dark brown with complete maculations distinctively orange to red-orange in color in western and central Mexico. The orange pigment of these maculations is apparently soluble in some solvents, and specimens "cleaned" of fat often have the orange replaced by a yellow-white color. Undersides coppery-purple except for the last few segments of the abdomen bright orange. The orange of the abdomen is exposed from above when the elytra are spread forward in flight.

Subspecies and morphological variants: An additional subspecies, *C. f. tenuilineata* Auduoin and Brullé, has been described from southeastern Mexico.

Map 59 Red-lined Tiger Beetle, *Cicindelidia fera.*

Distribution and habitats: One of the most common species in sandy and gravel banks of mountain streams of western and central Mexico, its regular range extends north to within 100 km of the Arizona border. A specimen found in the 1950s along a roadside ditch on the southern border of Arizona and New Mexico constitutes the only record for this species north of Mexico. If it arrived there on the winds of a summer storm or unintentionally on a vehicle from nearby Sonora is unknown. No established population of this species has been found in the United States after extensive search, and it should be considered an accidental member of the United States fauna. If it does occur anywhere regularly in the United States, it likely will be along desert canyon streams west of Nogales, Arizona.

Behavior: Runs quickly on sand, gravel and rocky stream beds and flies short distances when escaping danger.

Seasonality: It has been observed in every month of the year in Mexico, but the cold winters of southern Arizona and New Mexico probably narrow potential occurrence there to the summer or fall months.

Larval biology: Larva unknown.

Highlands Tiger Beetle, *Cicindelidia highlandensis* Choate
(Plate 17) [Map 60]

Description and similar species: Length 7–9 mm; above shiny black with green reflections and smooth elytral surface with no maculations. Below metallic purple and completely lacking white, hair-like setae (Fig. 4.15D). Abdomen brownish-red below and in flight the open elytra expose the brownish-red upper abdomen. Extremely similar to the Eastern Pinebarrens Tiger Beetle and the Scabrous Tiger Beetle, which occupy similar habitat. These latter two species are distinguished from the Highlands Tiger Beetle in the hand by the distinct pitting on the elytra surface, especially in the

Scabrous Tiger Beetle, a distinct row of white setae along the upper side of the thorax and the numerous white setae on the abdomen.

Subspecies and morphological variants: No distinct geographical populations of this extremely endemic species are known.

Map 60 Highlands Tiger Beetle, *Cicindelidia highlandensis.*

Distribution and habitats: Known only from open scrub pine, sandy ridges and fossil sand dunes of central Florida in Highland and Polk counties. Land development, citrus expansion, plant succession and fire suppression are the main threats to this highly endemic species, which is a candidate for official listing as endangered or threatened.

Behavior: Solitary or in small numbers among the scattered open patches of bare sand. Seeks shade along the edges of vegetated patches during the hottest part of the day. Easy to approach and a weak flier (escape flights only 3–6 m).

Seasonality: Adults active from May to early September with peak abundance from June to early July. It has a 1-year life cycle.

Larval biology: Larvae active throughout the year in shallow burrows (6–12 cm deep). Burrows usually in small clusters within bare patches of stabilized sand away from vegetation.

Eastern Pinebarrens Tiger Beetle, *Cicindelidia abdominalis* Fabricius (Plate 17) [Map 61]

Description and similar species: Length 8–11 mm; above shiny black with greenish-blue highlights and elytral surface with shallow pits and maculations reduced to dots for the middle maculation and a white tip to the end of the elytra. On some individuals, however, a white line along the rear edge of the elytra extends slightly forward. Below metallic blue with moderate density of white hair-like setae forming a row across the side of the thorax and covering parts of the abdomen (Fig. 4.15B). Abdomen brownish-red below and in flight the open elytra expose the brownish-red upper abdomen. The similar Highlands Tiger Beetle has a smooth elytral surface and no setae below or on the thorax, and the Scabrous Tiger Beetle has extensive pitting on the elytral surface and dense setae on the side of the thorax and on the abdomen.

Subspecies and morphological variants: No geographical populations have been described as subspecies.

Distribution and habitats: Paths, roads, and open bare sandy areas in pine barrens, sand hills, and scrub lands.

Behavior: Adults often hide near shaded areas and lichens in the middle of the day where their black color makes them look like pieces of dead wood or lichen. They are weak fliers and move short distances to escape danger. This Atlantic Coastal Plain species overlaps with the Scabrous Tiger Beetle only in northeastern Florida. It also occurs together with two other small tiger beetle species, the Moustached Tiger Beetle and the Whitish Tiger Beetle, both of which are lighter colored.

Map 61 Eastern Pinebarrens Tiger Beetle, *Cicindelidia abdominalis.*

Seasonality: Occurs from March to October, but most common in July and August. One- to 2-year life cycle.

Larval biology: Burrows in stabilized but deep sand and hard-packed soil.

Scabrous Tiger Beetle, *Cicindelidia scabrosa Schaupp* (Plate 17) [Map 62]

Description and similar species: Length 7–8 mm; above shiny black and elytral surface deeply pitted with maculations reduced to short white band and dot at tip of elytra. Below metallic purple with a dense band of setae on the side of the thorax and dense patches of white hair-like setae on the abdomen (Fig. 15.4F). Abdomen brownish-red below and in flight the open elytra expose the brownish-red upper abdomen. Distinguished from the Highlands Tiger Beetle, which has smooth elytra and no setae below or on the thorax, and from the Eastern Pinebarrens Tiger Beetle, which has shallow pitting on the elytral surface and moderately dense setae on the side of the thorax and on the abdomen.

Subspecies and morphological variants: Now considered to have no distinct subspecies.

Distribution and habitats: Confined to the Florida peninsula and adjacent southeastern Georgia, this species occupies small openings, roads, trails, dunes and edges in sandy pine forest, scrub lands, and tree plantations.

Behavior: Solitary and secretive hiding in grassy areas at the edge of open sandy areas. Difficult to find because of its small size and low densities. Flies short distances to escape danger. Becomes inactive during the hottest part of the day.

Map 62 Scabrous Tiger Beetle, *Cicindelidia scabrosa*.

Seasonality: Adults active from mid-May to September but most active from June to late August. One-year life cycle.

Larval biology: Burrows are found in the same habitats as the adults. Larvae are active all year around.

Miami Tiger Beetle, *Cicindelidia floridana* Cartwright (Plate 17) [Map 63]

Description and similar species: Length 7–8 mm; until recently this isolated population in the Miami area had been considered a subspecies of the Scabrous Tiger Beetle. Because of habitat destruction and urbanization, individuals of this species had not been observed in the wild since its initial discovery in the 1930's. However, in 2007 a small surviving population was

found in the Miami area. The combination of its dorsal coloration, distribution, and seasonality were used to determine it as a distinct species. It differs from the similar Scabrous Tiger Beetle in the shining dark green color of the head, thorax and elytra. The rear of the elytra have deep pits that reflect purple, and its maculations are reduced to a small white edge to the rear (apex) of each elytron.

Distribution and habitats: The Miami Tiger Beetle is separated from the range

Map 63 Miami Tiger Beetle, *Cicindelidia floridana*.

of the Scabrous Tiger Beetle by the marshy habitats of the Everglades and Lake Okeechobee in south central Florida. It is restricted to small sandy pockets of savanna-like forest on limestone outcrops (pine rockland habitat). This rare native habitat is maintained by fire and now found only in Miami-Dade County and parts of the Keys. The known population of this species survives in three small contiguous protected sites in the Richmond Heights area of Miami.

Behavior: Solitary and secretive, it hides in grassy areas at the edge of open sandy areas. Difficult to find because of its small size and low densities. Flies short distances to escape danger. Becomes inactive during the hottest part of the day.

Seasonality: Adults active from early May to mid-October.

Larval biology: Larvae likely occur in the same microhabitats with adults but are not described.

Limestone Tiger Beetle, *Cicindelidia politula LeConte* (Plate 17) [Map 64]

Description and similar species: Length 9–11 mm; highly variable in size and color; semi-isolated populations are metallic dark green, purple, blue, reddish or black above and maculations are absent or formed into a broad white band along the edge of the elytra. Below dark metallic purple, green or bronze. Abdomen brownish-red to orange below and in flight the open elytra expose the orangish upper abdomen. In central Texas it frequently co-occurs with Schaupp's Tiger Beetle on roads, trails, and ditches adjacent to limestone outcrops. It may be found with the Punctured Tiger Beetle, which lacks the orange abdomen, and the Eastern Red-bellied Tiger Beetle, which is dull above.

Subspecies and morphological variants: One relatively widespread subspecies and four restricted or semi-isolated subspecies have been described. Size, upperpart color and extent of maculations are the primary characters used to differentiate among them.

    C. politula politula **LeConte:** Above black to blue-black, elytral surface smooth and maculations absent or restricted to small white spot on the rear tip of each elytron. Below dark metallic purple except for contrasting orange abdomen. Most widely distributed subspecies; occurs from north central Mexico to south central Oklahoma (Carter and Murray counties).

    C. politula barbaraannae **Sumlin:** Above metallic red or brownish-red with a well-developed white line running along the edge of the elytra. Known only from above 1500 m in isolated mountains (Hueco Mountains, Sierra Diablo Mountains, and Apache Mountains) east of El Paso in extreme western Texas and in the Sacramento and Capitan Mountains of southern New Mexico.

C. politula laetipennis W. Horn: Above metallic purple, blue, or blue-green with a thin white line running along edge of the elytra or some individuals with maculations absent. Definitely known only from southeastern Coahuila, Mexico at 1750 m elevation, but further exploration may find that it occurs in southern Texas.

C. politula petrophila Sumlin: Above highly variable from blue or blue-green to greenish-black or even coppery-red; maculations usually absent, but some have partial maculations along the outer edge of the elytra. Known only from above 1670 m in the Guadalupe Mountains of west Texas and adjacent New Mexico.

C. politula viridimonticola Gage: Above bright green or bluish-green and rarely reddish; maculations consist of a broad white band along the outer edge of the elytra. Originally reported from a 1.5-hectare area in central Eddy County of southern New Mexico above 2193 m elevation, individuals matching this form have been found over a broader area in the Guadalupe Mountains National Park. Here they regularly occur together with individuals typical of *C. p. politula* and *barbaraannae*. Some experts do not feel this color form is a distinctively different group, and recent work has confirmed that the green form is not genetically distinctive from *barbaraannae*.

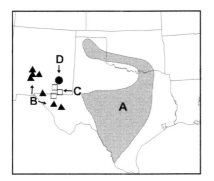

Map 64 Limestone Tiger Beetle, *Cicindelidia politula*;
A, *C. p. politula*; **B**, *C. p. barbaraannae*;
C, *C. p. petrophila*; **D**, *C. p. viridimonticola*.

Distribution and habitats: All populations of this species are tightly associated with limestone outcroppings, with adults found on gravel, rocks and large flat surface boulders of limestone. They may occur in open areas, low vegetation, or in forested sites at low elevations in the eastern portion of its range, but from 1500 to over 2500 m elevation in the west found in low, open vegetation; not found near water.

Behavior: The western subspecies are especially difficult to find and observe. They run quickly among the boulders and rocks, and if they fly, it is for only for a meter or so.

Seasonality: Because of their extremely dry habitat, rainfall is critical for adult activity. The various subspecies become active with the initiation of the summer rains,

July to October, but each local population is only active for three to eight weeks during their respective wet periods.

Larval biology: Larva unknown.

Cazier's Tiger Beetle, *Cicindelidia cazieri* Vogt
(Plate 18) [Map 65]

Description and similar species: Length 9–10 mm; above dull black with maculations reduced to small white dots and a white tip on the rear end of the elytra. In the hand, two parallel rows of shallow green pits can be seen along the inner edge of the elytra. Below dark metallic blue-purple. Abdomen brownish-red to orange below and in flight the open elytra expose the orangish upper abdomen. Most easily confused with nominate forms of the Limestone Tiger Beetle and the Eastern Red-bellied Tiger Beetle, but both of these species lack the parallel row of shallow green pits on the elytra, and the Limestone Tiger Beetle is shiny black above.

Subspecies and morphological variants: No distinct populations are known for this highly endemic species.

Map 65 Cazier's Tiger Beetle, *Cicindelidia cazieri*.

Distribution and habitats: Restricted to open scrub and thorn thickets with limestone gravel and sparse grass in the Rio Grande Valley of southern Texas (Jim Hogg and Starr counties) south into central Tamaulipas, Mexico.

Behavior: After fall and winter rains, often seen sunning on limestone slabs flush with the soil surface. Extremely wary and difficult to see against the multicolored gravel and vegetation.

Seasonality: Adults active following measurable rain in September and October. A few individuals can be active following rain in the spring.

Larval biology: Larva unknown.

Eastern Red-bellied Tiger Beetle, *Cicindelidia rufiventris* Dejean
(Plate 18) [Map 66]

Description and similar species: Length 9–12 mm; upperparts dull dark brown, black or dark blue with wide to thin, disrupted or no maculations.

141

Below metallic blue. Abdomen brownish-red to orange below and in flight the open elytra expose the orangish upper abdomen. In Texas most easily confused with nominate forms of the Limestone Tiger Beetle and Cazier's Tiger Beetle, but the Cazier's Tiger Beetle has a parallel row of shallow green pits on the elytra, and the Limestone Tiger Beetle is shiny black above. In the east the Twelve-spotted Tiger Beetle is similar and often in the same habitat as the Eastern Red-bellied Tiger Beetle. The Twelve-spotted Tiger Beetle, however, is more robust and lacks the orange abdomen. Also in the east, the nominate forms of the Punctured Tiger Beetle is confusingly similar, but it also lacks the orange abdomen.

Subspecies and morphological variants: Four subspecies have been described, and they are distinguished on the basis of upperpart color, extent of maculations and body size. All but *C. r. reducens* **W. Horn** of southwestern Mexico occur in the United States.

C. rufiventris rufiventris **Dejean:** Above dark brown to blackish with maculations broken into small spots and thin short lines. The nominate form occurs throughout much of the eastern United States but is absent from most of the Atlantic coastal plain in the southeast. A zone of intergradation with *cumatilis* runs through southern Missouri, Arkansas and Louisiana.

C. rufiventris cumatilis **LeConte:** Above dark blue with maculations usually absent or reduced to a small white spot at the tip of the rear end of the elytra. Occurs from Louisiana south through eastern Texas into northeastern Mexico but absent from coastal areas.

C. rufiventris hentzii **Dejean:** Above dark brown with complete or almost complete maculations. This highly endemic population is limited to moss-covered rocks, ledges, and granite quarries in the hills surrounding Boston, Massachusetts.

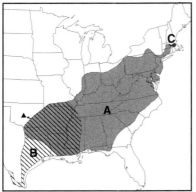

Distribution and habitats: Dry upland habitats with rocky outcroppings, gravel, sandstone, coarse sand, clay banks and sparsely vegetated clearings in open forest.

Behavior: Usually a solitary species, it sometimes occurs in large numbers. The patchy distribution of its habitat best explains its extensive but localized occurrence. It is a weak flier and makes short escape flights. They use shade for

Map 66 Eastern Red-bellied Tiger Beetle, *Cicindelidia rufiventris*; **A**, *C. r. rufiventris*; **B**, *C. r. cumatilis*; **C**, *C. r. hentzii*.

thermoregulation extensively in the morning and afternoon and cease most activity during the heat of mid-day. Commonly attracted to lights at night.

Seasonality: A summer active species, individuals can be found from May to September but most commonly in June and July. Overwinters as larvae and has a 1- or 2-year life cycle.

Larval biology: Burrows under rocks (*hentzii*) and in yellow clay (*rufiventris*).

Western Red-bellied Tiger Beetle, *Cicindelidia sedecimpunctata* Klug (Plate 18) [Map 67]

Description and similar species: Length 9–11 mm; above dark brown with wide but incomplete maculations. The middle band is distinctive as a complete wavy line that does not reach the edge of the elytra and is usually constricted in the middle. Below dark copper and blue with almost entire abdomen bright orange. In flight the spread elytra expose the bright orange upper abdomen. Within its range, most likely to be confused with three other species that are extremely similar and occur in the same habitats. The Ocellated Tiger Beetle can be distinguished by a middle maculation that is broken into two dots, and the orange on the abdomen is either absent or reduced to the last few segments. The Wetsalts Tiger Beetle is black to dark brown above, and its middle maculation usually reaches the edge of the elytra. Melissa's Tiger Beetle is restricted in the United States to Ponderosa Pine woodlands above 2500 m elevation in the Chiricahua Mountains of southeastern Arizona. It has the orange of its abdomen restricted to the last few segments, whereas the Western Red-bellied Tiger Beetle has an entirely orange abdomen.

Subspecies and morphological variants: Five subspecies are recognized for this species, from Costa Rica to Mexico, but only the nominate form extends into the southwestern United States.

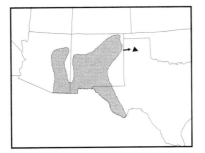

Distribution and habitats: It occurs along desert streams, lakes, temporary ponds, roadside ditches, irrigation canals, grassy marshlands, mud flats and alkali areas of the extreme southwest. Following summer rains it spreads to gravel paths and upland meadow areas at higher elevations up to 3000 m.

Behavior: In the early summer before the rains begin, highly gregarious with

Map 67 Western Red-bellied Tiger Beetle, *Cicindelidia sedecimpunctata*.

sometimes 1000s concentrated along the edge of water. As water sources dry up and tadpoles, insect pupae and other aquatic organisms become exposed, hundreds of Western Red-bellied Tiger Beetles will attack these large but helpless organisms and eat them. Following the onset of the first significant rain there apparently is a dispersal event as this species' populations numbers fall drastically in lowland areas just as individuals appear in adjacent higher elevation areas, where they occur on gravel paths and open meadows. Commonly attracted to lights at night.

Seasonality: Typically a summer species it is active earlier than most other desert species of tiger beetles in June and continues until September. Occasionally some adults will be active following spring rains in April and May.

Larval biology: Even though an abundant species, its larvae are unknown in the wild.

Melissa's Tiger Beetle, *Cicindelidia melissa* Duran and Roman (Plate 18) [Map 68]

Description and similar species: Length 9–11 mm; above dull brown with strong reddish and brassy reflections especially on the head and pronotum. Maculations are wide, but incomplete. The middle band usually consists of a wavy band that does not reach the edge of the elytra; other maculations are broken into dots. A row of blue-green or gold punctures is usually visible on each of the elytra. Below dark copper to olive brown and blue, with the last couple of segments orange to orange brown. It is very similar in appearance to the Western Red-bellied Tiger Beetle and may co-occur with this species at above 2500 m elevation in the Chiricahua Mountains of Arizona. The Western Red-Bellied Tiger Beetle, however, has an abdomen that is nearly all orange underneath, and is more uniformly brown on the head and pronotum.

Map 68 Melissa's Tiger Beetle, *Cicindelidia melissa*.

Subspecies and morphological variants: No distinct subpopulations are known for this species.

Distribution and habitats: This recently described species occurs most commonly in the mountains of western Chihuahua and eastern Sonora south to Durango in Mexico. It reaches the northern extreme of its distribution in the Chiricahua Mountains of southeastern Arizona in the United States. Found along upland rocky trails in Ponderosa Pine woodlands above 2500 m elevation.

Behavior: Generally found in small numbers, it occurs on sparsely vegetated upland trails usually away from water. It is wary and difficult to approach and when threatened will fly or run for cover in rock crevices or scrub vegetation.

Seasonality: Active from July to August. This species is only active after summer monsoon rains.

Larval biology: Larva unknown.

Ocellated Tiger Beetle, *Cicindelidia ocellata* Klug
(Plate 18) [Map 69]

Description and similar species: Length 9–13 mm; above brown to dark brown with maculations reduced to four spots on each elytron. Below metallic dark coppery-green. Abdomen completely dark metallic green or last few segments red-brown.

Subspecies and morphological variants: This largely Central American and Mexican species is divided into two subspecies, both of which enter the border states of the United States.

    C. ocellata ocellata **Klug:** Somewhat smaller and occurs in southeastern Arizona and adjacent New Mexico. The last few segments of its abdomen are orange.

    C. ocellata rectilatera **Chaudoir:** Occurs from New Mexico to Louisiana and has a larger body size and no orange on the abdomen.

Distribution and habitats: Moist open ground, such as edges of permanent and temporary ponds, irrigation ditches, stream banks, irrigated fields, moist pastures, alkali flats, and ocean beach. Following heavy rains expands into upland sites with sparse vegetation and bare soil patches. Rarely occurs above 2500 m elevation. Recently small populations have been found in Mississipi and Arkansas that may indicate an expansion of the species eastward.

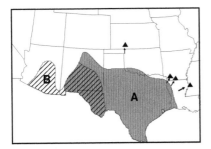

Map 69 Ocellated Tiger Beetle, *Cicindelidia ocellata*; **A**, *C. o. rectilatera*; **B**, *C. o. ocellata*.

Behavior: Often extremely abundant and gregarious around water edges but more solitary in upland habitats. Occasionally climbs vegetation to escape the heat of the soil surface at mid-day or to roost at night.

Seasonality: A summer active species, appears in the west following rains (June to October) and in the east from April to December. One-year life cycle.

Larval biology: Groups of burrows often in high densities in moist loam and clay in ditches, drainage areas and pond edges. Most of the burrows curve almost to the horizontal at their lower ends (12–25 cm deep).

Cobblestone Tiger Beetle, *Cicindelidia marginipennis* Dejean

Description and similar species: Length 11–14 mm; above dull olive with white band running along outer edge of elytra. Below dark metallic blue with bright red-orange abdomen whose color above is exposed in flight when the elytra are spread.

Subspecies and morphological variants: The separate population in Mississippi and Alabama tends to be larger and browner above than the northeastern populations, but no subspecies have been named.

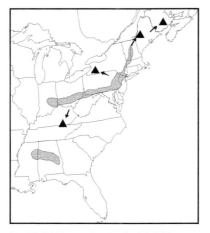

Map 70 Cobblestone Tiger Beetle, *Cicindelidia marginipennis.*

Distribution and habitats: Historically two disjunct populations were known, one in the northeastern United States and one in eastern Mississippi and central Alabama. Recently additional small populations were discovered in southeastern Kentucky, extreme southwestern New York state and southeastern New Brunswick, Canada. This species is restricted to wet pebbles, cobblestone, sand and sparsely vegetated bare areas on islands and along the edges of small to medium fast-flowing streams. It has apparently disappeared from many of its former sites because of dam construction, river channelization and water pollution.

Behavior: Very wary, this species is extremely well camouflaged and difficult to see. Occasionally gathers in large numbers in localized habitat patches on these river islands.

Seasonality: Adults active from May to September but most common in July.

Larval biology: Burrows are placed in wet sand among cobblestones of river islands.

Wetsalts Tiger Beetle, *Cicindelidia hemorrhagica* LeConte (Plate 18) [Map 71]

Description and similar species: Length 12–14 mm; above black, blue-green, to reddish-brown; maculations absent, interrupted or complete. The middle maculation, if present, is a line that generally does not reach the edge of the elytra in most populations. Below dark purple and copper with a bright orange abdomen. The bright orange upperpart of the abdomen is exposed when the elytra are spread in flight.

Subspecies and morphological variants: Four subspecies are presently recognized, but one of these, *C. h. hentziana* Leng, is known only from southern Baja California, Mexico.

C. hemorrhagica hemorrhagica LeConte: Above black to dark brown with three separate maculations of variable length, but usually wide and not connected to each other. The middle maculation, if present, often does not reach the edge of the elytra. The nominate form occurs throughout the western United States and intergrades so broadly with *C. h. woodgatei* in Arizona and New Mexico, it is difficult to separate many individuals. Along the coast of Southern California, a large proportion of the population has a distinctive blue sheen above with reduced or no maculations. This population is considered by some experts to be a separate subspecies, *C. h. pacifica* Schaupp.

C. hemorrhagica arizonae Wickham: Above reddish-brown with complete maculations that reach the outer edge of the elytra. Below coppery with orange abdomen. Surrounded on all sides by the nominate species, this population is restricted to the Colorado River at the bottom of the Grand Canyon in northern Arizona and along the Virgin River in adjacent Utah and Nevada.

C. hemorrhagica woodgatei Casey: Above black with reduced maculations that generally are thinner than those of the nominate subspecies. The reported range of this subspecies includes Arizona, New Mexico and western Texas, but the lack of consistent characters to separate it from the nominate form leaves its status in considerable doubt.

Distribution and habitats: Widely distributed across the western United States, the Wetsalts Tiger Beetle is rarely found away from water, in such

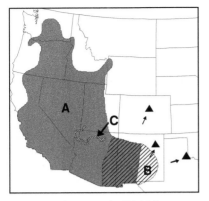

Map 71 Wetsalts Tiger Beetle, *Cicindelidia hemorrhagica*; A, *C. h. hemorrhagica*; B, *C. h. woodgatei*; C, *C. h. arizonae*.

sites as along edges of ponds, lakes, reservoirs, rivers, springs and irrigation ditches. It also occurs on salt flats, sea beaches, sandy estuaries and tidal flats along the ocean. Apparently it is capable of relatively long distance dispersal as there are records of vagrants reaching northern Texas and northern Colorado.

Behavior: Frequently occurs in high densities feeding on concentrated prey items in or along the edge of water. Occasionally but not regularly attracted to lights at night. It has relatively short escape flights.

Seasonality: Adults have been found active from April to October, but its peak of activity is dictated by the presence of water and summer rainfall, especially in the desert areas of its range. In most areas it is primarily a summer species from June to September.

Larval biology: Burrows are often densely aggregated in moist sandy clay at the edges of ponds, irrigation ditches, and saline lakes and springs.

Schaupp's Tiger Beetle, *Cicindelidia schauppii* G. H. Horn
(Plate 14) [Map 72]

Description and similar species: Length 8–11 mm; long, slender body; above brown with broad maculations usually connected in a band along the edge of the elytra and the middle maculations extending across the elytra to almost meet at the inner edges. Below metallic green with entire abdomen bright orange. Distinct within its range, it is, however, confusingly similar to some forms of the Variable Tiger Beetle. However, the two species do not overlap in distribution, and the Variable Tiger Beetle does not have a bright orange abdomen.

Subspecies and morphological variants: No distinct populations have been described, but populations on the coast of south Texas tend to have maculations more reduced than inland populations.

Distribution and habitats: Restricted to the southeastern Great Plains. It occurs in open, dry ground that is sparsely vegetated and with soil that is

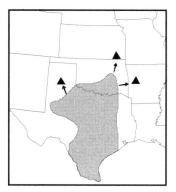

Map 72 Schaupp's Tiger Beetle, *Cicindelidia schauppii.*

usually saline limestone, gravel, or sand. Also found in baseball diamonds, roadside ditches, and parks.

Behavior: Attracted to lights at night.

Seasonality: A summer species in which individuals have been observed most regularly from August to October but occasionally in May. This species is usually dependent on rainfall for activity.

Larval biology: Burrows concentrated in large numbers in open weedy areas away from water.

Orange-spotted Tiger Beetle, *Cicindelidia sommeri* Mannerheim (Plate 16)

Description and similar species: Length 10–11 mm; above dark brown to blackish with three very wide orange maculations that generally reach the elytral edge but are not connected to each other. Below purplish-black with only a hint of orange on the abdomen. The orange pigment of these maculations present in central and northern Mexico populations is soluble in solvents. Thus "cleaned" specimens may acquire yellowish maculations.

Subspecies and morphological variants: No distinct subspecies are known.

Distribution and habitats: Occurring commonly in rocky, gravel and sand stream beds along the length of the Sierra Madre Occidental of west Mexico, this species reaches to within 150 km of Arizona and may eventually be found there. It is most likely to occur on continuations of small canyon streams that flow south from the United States west of Nogales. It is included here, however, on the basis of several specimens labeled as collected in San Diego County, California. Because this species has never been recorded to the south of California in Baja California, Mexico, there is considerable doubt about the reliability of these specimens' origin, and we consider the species hypothetical for the area covered in this field guide.

Behavior: Solitary individuals forage on wet sand among rocks and stones. They fly short distances to escape danger, but are relatively easy to approach.

Seasonality: In Sinaloa and southern Sonora, Mexico, this species is active from July to December.

Larval biology: Larva unknown.

S-banded Tiger Beetle, *Cicindelidia trifasciata* Fabricius
(Plate 14) [Map 73]

Description and similar species: Length 11–13 mm; above dull black, dark brown or olive. Maculations reduced and thin and often difficult to see in the field. However, if evident, the middle maculation with its sinuate "S" shape is diagnostic. Below dark purple to black.

Subspecies and morphological variants: Over its broad range from North America to the Caribbean and Chile there are eight recognized subspecies, of which only two occur in the United States. They are distinguished on the basis of their upper surface colors.

   C. trifasciata sigmoidea LeConte: Occurring on the southern Pacific coast of California, this subspecies is dark olive above with distinct maculations.

   C. trifasciata ascendens LeConte: Found broadly from the Atlantic coast and Gulf of Mexico to the Gulf of California it is blackish above with often thin or broken maculations.

Map 73 S-banded Tiger Beetle, *Cicindelidia trifasciata*; **A**, *C. t. ascendens* (darkly stippled = regular occurrence, lightly stippled = irregular but not unexpected, triangle with arrow = unexpected occurrence; **B**, *C. t. sigmoidea*.

Distribution and habitats: Most common in salt water-edge habitats close to the coast, it can be found in tidal mudflats, marshes, bays and inlets. It ranges inland along freshwater sources several hundred kilometers from the coast. It has also been found occasionally as an accidental disperser as far north as New England and as far inland as Kansas and Indiana, but no established populations have been recorded for these inland areas.

Behavior: Gregarious in warm coastal areas and extremely wary, this species has the ability to fly and/or use tail winds to accomplish very long flights. It has been recorded more than 160 km off shore on oil rigs in the Gulf of Mexico. Its normal escape flight is a vertical, horseshoe-shaped path, usually a meter or two high. Frequently attracted to lights at night.

Seasonality: Primarily a summer species active from June to August in the east and May to June in the west, a few individuals can be found active as early as April and as late as December. In Florida it is active every month of the year.

Larval biology: Burrows in sand of tidal flats just above the high tide line and at the edge of overwash areas. Burrow depth shallow (3–8 cm).

Saline Tiger Beetles, Genus *Eunota*

Most of the species in this group were recently redefined and moved from the *Habroscelimorpha* group based on DNA affinities. They are found primarily in semitropical and tropical areas of the Western Hemisphere. Several species are restricted to Mexico, South America and/or the Caribbean, and nine are found in the United States, chiefly in the southeast and southwest. The body shape tends to be elongated, the legs long, the eyes protruding and most species with expanded maculations. They are generally found on soils with high salt content.

California Tiger Beetle, *Eunota californica* Ménétriés
(Plate 19) [Map 74]

Description and similar species: Length 11–13 mm; above coppery-brown, dark green, or blue with maculations broadly connected along outer edge of elytra and a dark wedge separates the inner ends of the middle and rear maculations. Below dark metallic green and copper with last segments orangish below. Abdomen and thorax thickly covered with white hair-like setae, but no setae present on the cheeks of the head. Extremely similar to the Glittering Tiger Beetle and the Riparian Tiger Beetle but distinguished by the shape of the dark wedge between the inner ends of the middle and rear maculations on the rear part of the elytra (Fig. 4.25). In the California Tiger Beetle, the wedge extends farther backward than forward, but in the other two it extends the same distance both directions or farther forward than backward. Of these three similar species, only the California Tiger Beetle and the Riparian Tiger Beetle overlap geographically, and then only on the Salton Sea of southern California. In the hand, the Riparian Tiger Beetle also has distinct white hairs on its cheeks.

Subspecies and morphological variants: Four subspecies are recognized, but two of them are restricted to Mexico, **E. c. *californica* Ménétriés** on the coasts of central and southern Baja California, and **E. c. *brevihamata* W. Horn** on the coast of central Sonora and Sinaloa.

    E. californica mojavi **Cazier:** Above reddish-brown. Occurs from the Mojave Desert of inland Southern California south to the Salton Sea, Imperial County, and northwestern Sonora, Mexico. Intergrades with the subspecies *pseudoerronea* at Soda Lake in northeastern San Bernardino County, California.

    E. californica pseudoerronea **Rumpp:** Above dark green to blue. Restricted to Death Valley and the southern portion of the Armargosa Valley in eastern Inyo County, California.

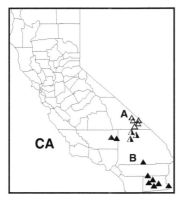

Distribution and habitats: Occurs away from vegetation on open wet, sandy or muddy saline soil, such as tidal flats, saline lakes, and muddy salt flats.

Behavior: Runs quickly over long distances. Wary and a strong flier. Commonly attracted to lights at night.

Seasonality: A summer species active from July to October in the south and from April to June in the northern part of its range.

Larval biology: Burrows (8–16 cm deep) are found in salt-encrusted sandy-clay soils along the edges of water and adjacent salt flats.

Map 74 California Tiger Beetle, *Eunota californica*; **A**, *E. c. pseudoerronea*; **B**, *E. c. mojavi.*

Western Tidal Flat Tiger Beetle, *Eunota gabbii* G. H. Horn
(Plate 19) [Map 75]

Description and similar species: Length 10–12 mm; above shiny reddish-olive with thin and distinctively elongated maculations that extend over most of the elytra. Below dark metallic green and densely covered with white, hair-like setae. Not readily confused with any other species.

Subspecies and morphological variants: No distinctive subspecies are known, but some individuals throughout its range have the maculations reduced or almost absent with dark reddish-olive elytra.

Distribution and habitats: Formerly along the southern California coast from Ventura to San Diego County and then south into Mexico. Now extirpated in the United States from all but three or four protected areas in Ventura, Orange and San Diego counties. Occurs on open wet saline soil with sparse vegetation, such as estuaries, tidal mud flats, salt marshes, and sea beaches.

Behavior: Often flies to land on the water to escape danger. Commonly attracted to lights at night.

Seasonality: Adults are active from July to September.

Map 75 Western Tidal Flat Tiger Beetle, *Eunota gabbii.*

Larval biology: Larva unknown.

Cream-edged Tiger Beetle, *Eunota circumpicta* LaFerté-Sénectère
(Plate 19) [Map 76]

Description and similar species: Length 12–14 mm; above blackish, copper, brown, dark green, or blue with a broad band of white along the edge of the elytra. Below dark metallic green to blue with thorax and sides of the abdomen moderately covered with white, hair-like setae. Only likely to be confused with the much smaller White-cloaked Tiger Beetle, which below has extremely dense white hair-like setae covering the entire surface including the head and cheeks.

Subspecies and morphological variants: Three subspecies are presently recognized, one coastal and two inland. Coastal forms are duller and more uniform in color while inland forms are brighter and tend to be highly variable in upperpart coloration. The elytral maculations tend to be wider in western populations.

   E. circumpicta circumpicta **LaFerté-Sénectère:** Above dark olive-green to dark purple with middle maculation extending inward from the elytral band at an oblique angle. Occurs along the Gulf Coast of Texas and inland along both the Rio Grande and into northeastern Texas. Intergrades with the subspecies *johnsonii* along the Oklahoma–Texas border.

   E. circumpicta johnsonii **Fitch:** Three color forms occur together in most populations; upperparts reddish, green, or blue with middle maculation

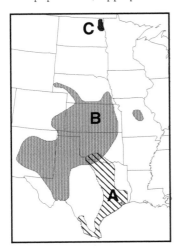

extending inward from the elytral band at right angle to the elytral edge. Occurs in the southern Great Plains with an isolated population in central Missouri. The Missouri populations are restricted to four salt springs and are exclusively greenish-blue above. Common at these springs as recently as the 1990s, their numbers have fallen off drastically due to habitat change.

   E. circumpicta pembina **Johnson:** Relatively small individuals that tend to be blackish or dark red above. Isolated by 800 km from the nearest population of the subspecies *johnsonii* to the south, this subspecies has a restricted range in North Dakota and perhaps extreme northwestern Minnesota in the area of the former Glacial Lake Agassiz basin.

Map 76 Cream-edged Tiger Beetle, *Eunota circumpicta*; **A**, *E. c. circumpicta*; **B**, *E. c. johnsonii*; **C**, *E. c. pembina*.

Distribution and habitats: Found on moist, muddy, saline flats along streams, ponds, lakes, drainage ditches, sea beaches, estuaries, and tidal flats.

Behavior: Often gregarious and a fast runner; usually ceases activity during the middle of the day and hides under cover in holes, vegetation, and detritus. Frequently attracted to lights at night.

Seasonality: Adults are active from May to October with peak abundance in June and July. Overwinters as larvae.

Larval biology: Burrows are found in open saline flats or near sparse vegetation. Burrow depth is 12–30 cm.

Riparian Tiger Beetle, *Eunota praetextata* LeConte
(Plate 19) [Map 77]

Description and similar species: Length 11–14 mm; upperparts reddish-brown with maculations connected at their bases into a continuous band. Middle maculation extending inward at an oblique angle on the elytra and almost touching the inner portion of the rear maculation. A wedge of dark color separates the two rear maculations and this wedge extends as far forward or farther than it does backward (Fig. 4.25). Below coppery-purple with moderate covering of white, hair-like setae on thorax and sides of abdomen. Extremely similar to reddish-brown forms of California Tiger Beetle and Glittering Tiger Beetle but separated geographically and by habitat from both except at the Salton Sea, where it occurs with the California Tiger Beetle. Both the California Tiger Beetle and Glittering Tiger Beetle lack obvious white setae on the cheeks, which are present in the Riparian Tiger Beetle (Fig. 4.23). In addition, the dark wedge of color on the rear of the elytra in the California Tiger Beetle extends farther backward than forward.

Subspecies and morphological variants: Two subspecies are recognized and distinguished by leg color and extent of maculations.

E. praetextata praetextata LeConte: Found on the lower Colorado, Gila and Salt Rivers and Salton Sea, it has completely metallic green legs.

E. praetextata pallidofemora Acciavatti: Restricted to the Virgin River of southern Nevada and adjacent Utah, it has yellowish upper legs as well as more diffuse maculations.

Distribution and habitats: Restricted to sandy beaches of desert rivers in the extreme southwest United States, as well as in the Salton Sea basin. Now absent from most parts of the Gila and Salt Rivers that no longer flow for much of

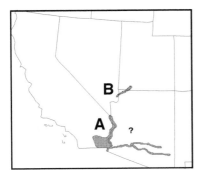

Map 77 Riparian Tiger Beetle, *Eunota praetextata*; **A**, *E. p. praetextata*; **B**, *E. p. pallidofemora*.

the year because of dams upriver. Recent habitat recovery and river restoration of stretches of these rivers in central Arizona have enabled the Riparian Tiger Beetle to recolonize from remnant river refugia.

Behavior: Adults move into grassy borders of desert river beaches during the middle of the day and often occur in large numbers. Commonly attracted to lights at night.

Seasonality: Adults active from May to August in the southern part of the range and June to August in the north.

Larval biology: Larva unknown.

Glittering Tiger Beetle, *Eunota fulgoris* Casey
(Plate 19) [Map 78]

Description and similar species: Length 11–14 mm; above coppery-red, brown or dark green to blue with maculations connected at their bases into a continuous band. Middle maculation extending inward at an oblique angle on the elytra and almost touching the inner portion of the rear maculation. A wedge of dark color separates the middle and rear maculations and this wedge extends as far forward or farther then it does backward (Fig. 4.22). In some populations the maculations are so expanded and coalesced they cover all but a central section of the elytra. Below dark metallic green to green-blue and moderately covered with white hair-like setae on the thorax and sides of the abdomen but absent from the cheeks. Similar to the Riparian Tiger Beetle and the California Tiger Beetle, but does not overlap geographically with either of these species.

Subspecies and morphological variants: Three subspecies are recognized, and they are separated by the color of their upperparts and the extent of maculation coalescence on the elytra.

E. fulgoris fulgoris **Casey:** Above coppery red with a distinct white band running along the edge of the elytra. Occurs along the Little Colorado River of northeastern Arizona and upper Rio Grande basin of New Mexico.

E. fulgoris erronea **Vaurie:** Above dark green to blue with a narrow white band running along the edge of the elytra. Isolated population restricted to the northern end of the Sulphur Springs Valley, Cochise County, in south-eastern Arizona.

E. fulgoris albilata Acciavatti: Above coppery-brown with white maculations extremely expanded and confluent—covering most of the elytral surface. Known from the Salt Basin of west Texas and adjacent New Mexico as well as several sites in the southern panhandle of Texas. Intergrades with nominate populations in south central New Mexico.

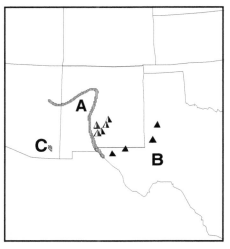

Map 78 Glittering Tiger Beetle, *Eunota fulgoris*; **A**, *E. f. fulgoris*; **B**, *E. f. albilata*; **C**, *E. f. erronea*.

Distribution and habitats: Restricted to muddy salt flats bordering ponds, lakes, drainage ditches and moist fields.

Behavior: Regularly attracted to lights at night.

Seasonality: A summer species, but activity of local populations is influenced by the summer rains—August to October in west Texas, July to September in southeastern Arizona, and June to September elsewhere.

Larval biology: Burrows are in open flats and small mounds along the edges of ponds and beaches. Burrow depths are 18–23 cm. Larvae are heavily attacked by bee-fly parasitoids in the family Bombyliidae (Fig. 7.2).

Gulfshore Tiger Beetle, *Eunota pamphila* LeConte
(Plate 20) [Map 79]

Description and similar species: Length 11–12 mm; above shiny green-olive or bronze with maculations coalesced to form a broad band along the elytral edge and a middle maculation that extends obliquely inward to end in the shape of a barbed harpoon point. Below dark bronzy-green with dense white, hair-like setae on the sides of the thorax and abdomen. Hind legs very long.

Subspecies and morphological variants: No distinct subspecies are known.

Distribution and habitats: Confined to bare, muddy, saline soil on sea beaches, tidal flats, and salt marshes along the Gulf of Mexico coast south to Tamaulipas,

Map 79 Gulfshore Tiger Beetle, *Eunota pamphila.*

Mexico. Regular as far east as the coast of Alabama.

Behavior: Usually becomes active early in the morning. Can be found at night by walking through its habitat with a flashlight. It may run around in circles in the beam permitting the observer to get close. Locally abundant and a weak flier.

Seasonality: Adults active May to December.

Larval biology: Larva unknown.

Saltmarsh Tiger Beetle, *Eunota severa* LaFerté-Sénectère (Plate 20) [Map 80]

Description and similar species: Length 12–15 mm; above shiny black, olive or dark green with maculations reduced to a spot on the middle edge of the elytra and a short "J"-shaped maculation at the rear end of the elytra. Below blackish to dark green with dense white hair-like setae on the sides of the thorax and abdomen. Labrum ivory-white. Similar to the Punctured Tiger Beetle but the Saltmarsh Tiger Beetle lacks the pits in the elytra and occurs in saline muddy areas not dry upland. Most easily confused with the closely related Elusive Tiger Beetle, which has two rows of obvious green pits down the inner edges of the elytra, sparse and short setae on the underside, and dark brown (males) to black (females) upper lip (labrum).

Subspecies and morphological variants: The nominate subspecies occurs in the United States south into eastern Mexico. Green morphs tend to predominate in some populations of Florida and black predominates in Texas, but not consistently enough to separate them into a subspecies. An additional subspecies, **E. s. *yucatana* W. Horn**, is endemic to the Yucatan Peninsula of Mexico.

Distribution and habitats: Occurs in open ground with moist mud or sand with sparse, low vegetation, such as salt marshes, coastal salt flats, sea beaches, lagoon edges, and drying

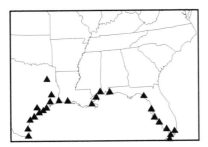

Map 80 Saltmarsh Tiger Beetle, *Eunota severa.*

lake beds. Found mainly on the Gulf of Mexico coast but occasionally inland a few kilometers on brackish or fresh water mud flats, but also at saline areas of northeastern Texas, 150 km from the coast.

Behavior: Most active in the morning and evening. Readily attracted to lights at night. Quite wary with a long escape flight. Often found together with the Gulfshore Tiger Beetle and the Coastal Tiger Beetle. During the hottest part of the day, adults may hide in vegetation near water and may be flushed out by walking along the edge of the vegetation. Adults will frequently enter fiddler crab holes to escape capture. Regularly scavenges dead fish and other organisms.

Seasonality: Adults active from June to September.

Larval biology: Larva unknown.

Elusive Tiger Beetle, *Eunota striga* LeConte
(Plate 20) [Map 81]

Description and similar species: Length 11–17 mm; above shiny black to dark green with maculations reduced to small spots in the middle and end of the elytra. A series of shallow pits runs along the inner edge of each elytron. Below blackish with sparse and short hair-like setae on the sides of the thorax and abdomen. Labrum black (females) or dark brown (males). Very similar to and co-occurs with the Saltmarsh Tiger Beetle, which lacks the rows of shallow pits on the elytra, has much more dense white setae on the underside, and a white labrum.

Subspecies and morphological variants: No distinct geographical populations are known.

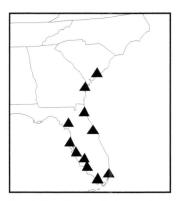

Distribution and habitats: Specific habitat not well known, but has been collected from salt marsh areas, salt mud flats and meadow-like openings in coastal pine forest along the west and northeast coasts of Florida to South Carolina as well as from inland Florida.

Behavior: Active primarily at twilight and at night. It has only been collected around lights at night. Apparently no one has ever observed it during the daytime.

Seasonality: Adults are found at lights from June to August.

Map 81 Elusive Tiger Beetle, *Eunota striga*.

Larval biology: Larva unknown.

White-cloaked Tiger Beetle, *Eunota togata* LaFerté-Sénectère (Plate 20) [Map 82]

Description and similar species: Length 9–14 mm; above coppery-red or brown with maculations formed into a white band along the edge of the elytra or expanded to almost cover the entire elytra. Below coppery but almost completely covered with dense white, hair-like setae on the head, thorax and abdomen. The spine at the rear end of the inner edge of the elytra is at the tip of the elytra in males and positioned forward away from the tip on the females to varying degrees among populations. Similar to the White-sand Tiger Beetle, which is restricted to inland rivers of the southeast, and to the Cream-edged Tiger Beetle, which is generally larger and lacks the thick coat of white, hair-like setae that cover the underside and head of the White-cloaked Tiger Beetle.

Subspecies and morphological variants: Three subspecies have been named, and they are separated on the basis of their color above, degree of maculation coalescence, and the degree to which the elytral spine on females is located forward and away from the rear tip of the elytra (retraction). Larger individuals occur in west Texas.

E. togata togata **LaFerté-Sénectère:** Above brown, and the white band along the elytral edge relatively narrow. The female elytral spine is greatly retracted. The nominate race occurs along the Gulf Coast from northwest Florida (Dixie County) to southern Mexico. Vagrant individuals and temporary colonies have also been recorded from coastal South Carolina, but the species has not been able to establish itself there.

E. togata globicollis **Casey:** Upperparts coppery with the white band along the edge of the elytra broad. The female elytral spine is only slightly retracted. This subspecies occurs in the central and southern Great Plains.

E. togata fascinans **Casey:** The maculations of the elytra are so greatly expanded they cover virtually the entire elytra. This subspecies is known only from salt flats in central New Mexico (Torrance County) and west Texas (Hudspeth County). There is a narrow zone of intergradation to the east with the subspecies *globicollis*.

Distribution and habitats: Occurs in the southern Great Plains and Gulf Coast where it is restricted to damp alkali and salt flats, roadside ditches, banks, abandoned oil fields and sand bars along rivers, shores of lakes and ponds with sparse vegetation. In coastal

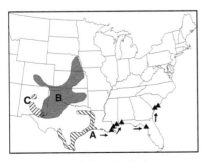

Map 82 White-cloaked Tiger Beetle, *Eunota togata*; **A,** *E. t. togata*; **B,** *E. t. globicollis*; **C,** *E. t. fascinans*.

areas occurs along sea beaches, lagoons, inlets, estuaries, upper areas of marshes, ponds and coastal lakes.

Behavior: Active both day and night and readily attracted to lights at night. Wary and difficult to approach. Tends to fly close to the ground. Apparently capable of long distance dispersion in coastal areas as a colony appeared for a few years in the 1930s in coastal South Carolina, 500 km to the east of the nearest known permanent part of its range in the panhandle of west Florida. Extremely well adapted for activity on hot substrates even in the middle of the day. Often occurs together with the Cream-edged Tiger Beetle and the Nevada Tiger Beetle.

Seasonality: Adults active from April to October, but most common August to September in the west, July in the Central Plains and May to August on the Gulf Coast.

Larval biology: Burrows are found mainly in saline clay or sandy-clay soils with poor drainage. Burrow depth is 10–18 cm.

Coral Beach Tiger Beetles, Genus *Microthylax*

The three species included in this genus are restricted to the Greater Antilles of the Caribbean and coastal western Mexico. They share an elongated thorax and greatly reduced elements of the male aedeagus and internal sac. Only one species enters the United States.

Olive Tiger Beetle, *Microthylax olivacea* Chaudoir
(Plate 25) [Map 83]

FL

Map 83 Olive Tiger Beetle, *Microthylax olivacea*.

Description and similar species: Length 10–12 mm; above shiny olive with three distinct maculations that are not connected to each other. Below coppery-green to purple.

Subspecies and morphological variants: No distinct subspecies have been described.

Distribution and habitats: Endemic to the coast of Cuba, but in the United States historically recorded from extreme southern Florida on the Gulf of Mexico sides of the outermost Florida Keys. Here it occupied open rocky soil, lithified coral, coarse sand and broken shells. Apparently it was a recent colonist, probably arriving with a hurricane from Cuba in the 1940s, but it has not been reported at any site in the Keys since the 1980s.

Behavior: Adults are found primarily on dark rocks at the edge of the ocean. They fly erratically to escape danger, but soon return to these dark rocks on which they are well camouflaged. Attracted to lights at night.

Seasonality: Adults active on the Florida Keys from June to August.

Larval biology: Larva is unknown.

Habro Tiger Beetles, Genus *Habroscelimorpha*

The species of this genus are all confined to ocean beaches. They are characterized by extremely long legs in both sexes. Only one species occurs in the United States. Recent DNA analysis shows that most of the species previously considered *Habroscelimorpha* belong in the genus *Eunota*.

Eastern Beach Tiger Beetle, *Habroscelimorpha dorsalis* Say (Plate 20) [Map 84]

Description and similar species: Length 8–15 mm; above bronze to greenish with extensive maculations that run the length of the elytra and are coalesced in some populations to cover most of the elytral surface. Abrasion by sand makes elytra of older individuals lighter. Below dark bronze to dark green with dense white hair-like setae covering the sides of the abdomen. Last pair of legs exceptionally long. Males and females are noticeably different in the shape of the thorax (cylindrical in males, trapezoidal in females), and the shape of the elytral rear tip (rounded in males, broadly notched in females).

Subspecies and morphological variants: Five subspecies have been described, and they are separated on the basis of body size and extent of maculation coalescence. Molecular DNA data demonstrate that the Atlantic populations are markedly distinct from the Gulf populations. However, the distinctions between the two subspecies from the Atlantic coast are not supported by these genetic data. All but one subspecies are found in the United States. The subspecies **H. d. castissima Bates** is restricted to Cuba and the Caribbean.

H. dorsalis dorsalis Say: The largest subspecies (13–15 mm), the nominate form has wide, cream-colored maculations that are frequently expanded to cover much of the elytral surface. This subspecies formally occurred on ocean beaches from Cape Cod, Massachusetts south to the Chesapeake Bay. The mean body size decreases and the elytra become darker from north to south. Due to recreational beach development, destruction of habitat by vehicles, and stabilization of beaches by artificial jetties and bulkheads, the present distribution of the nominate population has been reduced to two sites on the coast of Martha's Vineyard in Massachusetts and along both shores of the Chesapeake Bay in Maryland and Virginia. **NOTE:** This subspecies is federally listed in the United States as THREATENED and is illegal to collect.

H. dorsalis media **LeConte:** The second largest subspecies (11–14 mm), *media* has moderately wide maculations with distinct dark areas of bronze-green exposed. It occurs along the Atlantic Coast from southern New Jersey to southern Florida. There is little historical intergradation where it overlaps with the nominate subspecies. It has disappeared from about 20% to 30% of its former habitat, most likely due to heavy vehicle and recreational use of beaches.

H. dorsalis saulcyi **Guérin-Méneville:** Most members of this subspecies (10–12 mm) have the maculations completely expanded to cover all but the inner edges of the elytra in pure white. It occurs along the west coast of Florida to the Mississippi River delta in Louisiana. There it intergrades with the smaller and darker *venusta*. A gap of almost 250 km at the tip of the Florida Peninsula where there are no sandy beaches separates *media* from *saulcyi* populations.

H. dorsalis venusta **LaFerté-Sénectère:** This smallest subspecies (8–11 mm) has reduced maculations that make *venusta* the darkest population in the United States. It occurs from the coastal islands of Louisiana south along the Gulf Coast to Texas and northeastern Mexico.

Distribution and habitats: The Eastern Beach Tiger Beetle is an Atlantic and Gulf Coast species generally restricted to wide and dynamic sandy ocean or bay beaches with dunes or cliffs on the upper beach from south Texas to Massachusetts.

Behavior: Adults forage both day and night along the intertidal zone for amphipods and small insects. They also often scavenge dead fish and invertebrates. Mating occurs in the late afternoon and at night. Often highly gregarious on beaches with little human disturbance. Escape

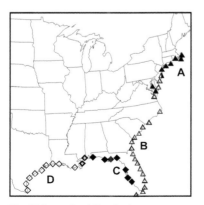

Map 84 Eastern Beach Tiger Beetle, *Habroscelimorpha dorsalis*; **A**, *H. d. dorsalis*; **B**, *H. d. media*; **C**, *H. d. saulcyi*; **D**, *H. d. venusta*.

flight is usually for several meters landward from the shoreline. Frequently attracted to lights at night.

Seasonality: Adults active from April to September, but mostly July and August, rarely to October. Overwinters as larvae and has a 1- to 2-year life cycle.

Larval biology: Burrows are 15–50 cm deep and are found from the high tide line to the base of sand dunes above the beach. In northern parts of the range, the larvae move their burrows up the beach in October to avoid winter storm floods and narrowing of the winter beach. Larvae are active primarily at night and plug the entrance to their burrows during warm days when the sand surface dries out. Recently larvae of the *media* subspecies were photographed looping their body and rolling like a wheel driven by the wind across sandy beaches of Georgia (Fig 7.3). They apparently use this behavior to escape predators when removed from their shallow tunnels, and they can travel up to 60 m across smooth sand at 3 m per second.

Opilid Tiger Beetles, Genus *Opilidia*

The six species of this genus are all confined to ocean beaches from Mexico to northern South America. They are characterized by males having an extremely small aedeagus, and both sexes have extremely long legs. Only one species has occurred in the United States, and then apparently as an isolated colony early in the last century.

Lime-headed Tiger Beetle, *Opilidia chlorocephala* Chevrolat (Plate 21) [Map 85]

Description and similar species: Length 6–8 mm; above olive-brown with wide white maculations connected along the edge of the elytra. Below dark bronzy with thick hair-like white setae on the sides of the thorax and front part of the abdomen.

Subspecies and morphological variants: Two subspecies are known, but only *O. c. smythi* E. D. Harris, has been found in the United States. The nominate form is larger with thinner maculations and occurs along the central Gulf Coast of Mexico south to Veracruz.

Distribution and habitats: More than 80 specimens of *smythi* were collected near Brownsville on exposed sand beaches of the ocean side of South Padre Island in south Texas in June of 1912. They are the only known specimens for the subspecies *smythi*, and no additional specimens of this subspecies

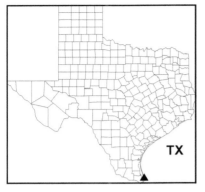

Map 85 Lime-headed Tiger Beetle, *Opilidia chlorocephala*.

have been collected in Texas or in adjacent Mexico. It remains a mystery as to what happened to this population. However, an unusually active and extended period of hurricanes hit the south Texas coast beginning in 1910, and then again in 1912, 1913, 1915, and an especially huge and destructive one struck in 1919. If the larvae and adults of this subspecies had a restricted range and a very specific habitat, it could have been wiped out completely shortly after its discovery by beach erosion and the short time for natural habitat recovery processes between hurricanes during this period of exceptionably stormy years.

Behavior: Extremely fast runner across wet sand at the ocean's edge. Found together with *venusta* subspecies of the Eastern Beach Tiger Beetle.

Seasonality: The Texas specimens were collected in June; adults in Mexico are active in June and July.

Larval biology: Larva unknown.

Little Tiger Beetles, Genus *Brasiella*

The nearly 40 species of this genus are concentrated in Central America, the Caribbean and South America. Only two species extend north to barely cross the Mexican border into the southwest United States. Another species is included for the United States as a vagrant from Cuba to Florida. The members of this genus are characterized by extremely small body size, the tip of the male aedeagus ending in a harpoon-like or crochet hook tip, and by peculiarities in the male internal genitalic sac.

Cuban Green-necked Tiger Beetle, *Brasiella viridicollis* Dejean
(Plate 21) [Map 86]

Description and similar species: Length 8 mm; above, head and thorax bright metallic green and elytra reddish-brown covered with small metallic green spots, which are best visible under a microscope. The maculations are thin and incomplete. Legs yellowish orange. Below dark metallic blue-black. The only other tiger beetle species found in the United States that are this small have the color of the head and thorax the same as the elytra, reddish-green to reddish-brown (Sonoran Tiger Beetle), or dark brown (Pygmy Tiger Beetle, Ant-like Tiger Beetle, and Swift Tiger Beetle).

Subspecies and morphological variants: No distinctive subspecies have been described.

Map 86 Cuban Green-necked Tiger Beetle, *Brasiella viridicollis.*

Distribution and habitats: Patches of bare ground in dry, upland grassy areas of the interior of southwestern (Piñar del Rio) to eastern Cuba. A male specimen of this species was collected at a black light on 4 June 1983 on Sugarloaf Key in the Florida Keys by W. H. Cross. This individual was most likely a vagrant carried from Cuba, perhaps by a storm. However, from 1979 (Hurricane "David") to 1992 (Hurricane "Andrew") a 13-year period of hurricane inactivity saw no major meteorological events that could easily explain the appearance of this specimen in Florida.

Behavior: Runs along paths through grassy fields and flies weakly.

Seasonality: Recorded active during the wettest months in Cuba from May to September.

Larval biology: Larva unknown.

Sonoran Tiger Beetle, *Brasiella wickhami* W. Horn (Plate 21) [Map 87]

Description and similar species: Length 7–8 mm; above shiny copper often with greenish reflections. Maculations incomplete and often broken into short lines and dots. Below dark metallic blue. Legs light brown with coppery reflections. The only other small species that occurs in the southwest is the Pygmy Tiger Beetle, but it is dark brown with green shallow pits in the elytra and has dark, metallic-colored legs. The only other coppery-red species in the southwest is the White-striped Tiger Beetle, but it is larger and has a bold yellowish-white stripe running the length of each elytron.

Subspecies and morphological variants: No distinct populations have been described.

Map 87 Sonoran Tiger Beetle, *Brasiella wickhami.*

Distribution and habitats: Found in open muddy areas with sparse grass, such as small temporary ponds and irrigation ditches.

Behavior: Runs fast along water's edge or across small muddy areas, and flies short distances to escape danger. Often attracted to lights at night.

Seasonality: Activity initiated by summer rains from June to August.

Larval biology: Larva unknown.

Pygmy Tiger Beetle, *Brasiella viridisticta* Bates
(Plate 21) [Map 88]

Description and similar species: Length 7–9 mm; above dark brown with distinct green shallow pits on the elytral surface. Maculations reduced to dots at the middle maculation and a short "J"-shaped line at the end of the elytra. Below dark metallic purple to blue-green. Recent DNA analysis places this species in the genus *Brasiella* rather than its previous position within the *Parvindela* (formerly *Cylindera*). The only other tiger beetle of such small size in the southwestern United States, the Sonoran Tiger Beetle, is reddish-copper above.

Subspecies and morphological variants: Three subspecies of this primarily Mexican species have been described, distinguished by the presence and pattern of shallow pits on the elytra. Only *B. v. arizonenesis* **Bates** enters the United States, in southeastern Arizona and southwestern New Mexico.

Map 88 Pygmy Tiger Beetle, *Brasiella viridisticta.*

Distribution and habitats: Open ground among vegetation at edges of permanent ponds, river beds, irrigation ditches, shaded trails in open forest, paths near streams, and gardens in towns and cities.

Behavior: Hides among vegetation to escape danger and runs quickly from one hiding spot to the next. Flies weakly if pressed but then for only a meter or less. Often found together with the

Western Red-bellied Tiger Beetle and the Ocellated Tiger Beetle. Occasionally attracted to lights at night.

Seasonality: Adults active from July to October. Overwinters only as larvae and the life cycle is 1 year.

Larval biology: Burrows in moist clay or sandy-clay soils in water edge habitats.

American Diminutive Tiger Beetles, Genus *Parvindela*

Based on DNA analysis, this genus recently was erected as a split from the Old World genus *Cylindera*. It is endemic to North America and its members superficially resemble the *Cylindera* by having a rounded thorax with few or no setae (glabrous). They are small in size and many are weak fliers or flightless. Species of *Parvindela* are active as adults only in the summer.

Variable Tiger Beetle, *Parvindela terricola* Say
(Plate 22) [Map 89]

Description and similar species: Length 8–12 mm; highly variable, dull brown, reddish-brown, green, blue or black above. The maculations are either absent, thin and unconnected, or broadly connected along or above the outer elytral edge. Body shape long and thin with relatively straight and parallel outer elytral edges. Below metallic green-blue with sparse white, hair-like setae on the sides of the thorax and abdomen.

Subspecies and morphological variants: At least 12 names have been proposed for various populations of this complex of color forms. It may include several species, each with its own array of subspecies, and some of which apparently overlap in distribution. Numerous attempts to make clearer taxonomic sense of this group using classic adult characters as well as molecular DNA have yet to be successful. A better knowledge of the distribution of forms, zones of intergradation, and detailed studies of genes and genitalia will be needed before the species and subspecies involved can be more reliably presented. Because these detailed studies are not yet available, we provide descriptions of the forms and names of the most distinct populations with the understanding that subsequent evaluations may alter their status.

P. terricola terricola Say: Above black, blue or dark brown with maculations absent in the eastern part of its range in northwestern Minnesota, North Dakota and adjacent parts of Manitoba and more complete maculations that are often connected to each other in the area of intergradation with

P. t. cinctipennis in the western prairies. Blue forms are common in some areas. The population in the northeastern part of its range has the surface of the elytra distinctly pitted, and it is considered a separate form, *P. pusilla* Say, by some experts.

P. terricola cinctipennis LeConte: Above brown, brownish-green, green, or blue with complete maculations that are joined along the edge of the elytra. Found east of the Rocky Mountains in lower elevations from Yukon to Colorado, New Mexico and Arizona.

P. terricola continua Pearson, Knisley and Kazilek: This stocky and large form (10–11 mm) is dark brown to green-brown above with all three maculations usually connected above the elytral edge until reaching the edge at the rear tip of the elytra. It is restricted to interior Southern California in Ventura, Kern and adjacent San Bernardino County and an isolated population in southern Nye County, Nevada that is primarily blue to purple.

P. terricola imperfecta LeConte: Above usually dark brown and blue or brown-green with first two thin maculations joined above the elytral edge. The long, branching middle maculation is diagnostic. Ranges east of the Cascades and Sierra Nevada across the Great Basin from southeastern British Columbia to southwestern Utah and northeastern California. Isolated populations west of the Cascades in the Puget Sound area of Washington.

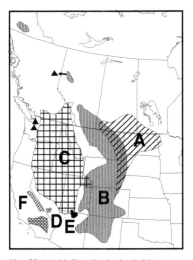

P. terricola kaibabensis W. Johnson: Above copper-green to blue-green, brown or red-brown with narrow maculations sometimes connected. This population was described from grassy meadows on the Kaibab Plateau of northern Arizona and southern Utah, but it is so variable and indistinct that its status as a subspecies is highly questionable.

P. terricola susanagreae Kippenhan: This all black to dark brown population is confined to the Owens Valley of eastern California (Inyo County)

Map 89 Variable Tiger Beetle, *Parvindela terricola*; **A**, *P. t. terricola*; **B**, *P. t. cinctipennis*; **C**, *P. t. imperfecta*; **D**, *P. t. continua*; **E**, *P. t. kaibabensis*; **F**, *P. t. susanagreae*.

Distribution and habitats: Widely distributed in the west, the Variable Tiger Beetle occurs on open grassy ground that is muddy or dry but usually alkaline or salty. Many populations are found in grassy areas of dry creek beds, salt lakes, irrigation ditches, banks and bars along rivers, sage brush areas, and shaded paths in open forest. Populations at higher elevations need moist soil along the edge of meadows.

Behavior: Usually solitary, it uses fast running into and among grass to escape predators but can also fly short distances. When it flies, the flight is erratic, usually landing in a clump of grass.

Seasonality: Over most of its range, the Variable Tiger Beetle is active as adults from April to September but most common in June to July, lingering into early September at higher elevations.

Larval biology: Burrows have been found among bare patches of open clay soils.

Meadow Tiger Beetle, *Parvindela lunalonga* Schaupp (Plate 22) [Map 90]

Description and similar species: Length 8–11 mm; above black or brown, rarely with green to blue on head and thorax. Maculations range from none to three broad bands partially connected. Recently elevated to a full species, it was formerly considered a subspecies of the similar Variable Tiger Beetle and is best distinguished by geographical range and the shape of the middle band if present. In the Meadow Tiger Beetle a complete middle band originates at a right angle to the outer edge of the elytra. The inside portion of the rear band is noticeably expanded.

Subspecies and morphological variants: No subspecies are recognized, but there is elevational variation in maculation patterns and color of the upperparts. Low in the San Joaquin Valley, most individuals are dark

brown with no green reflections on the elytra, and they exhibit the entire range of maculations from none to complete. At higher elevations along the west slope of the Sierra Nevada, individuals tend to be dark brown above with broader maculations and distinct green reflections on the elytra.

Distribution and habitats: This species occupies two different habitats within California. At low elevations it is around moist alkali habitats within the San Joaquin Valley, and at higher elevations in the central and northern Sierra Nevada, it is in moist mountain meadows. Specimens

Map 90 Meadow Tiger Beetle, *Parvindela lunalonga*.

of this species reported from the San Pedro Martir Mountains of northern Baja California in Mexico may be misidentified. This species has been extirpated from most of its range and is presently known from only one site in northern California and several newly discovered populations associated with irrigation ditches west of Stockton.

Behavior: Adults run erratically on bare patches of soil and escape danger by either running into low vegetation on the margins or by short flights to nearby patches of bare soil.

Seasonality: At lower elevations in the San Joaquin Valley, adults are active from May to July. At higher elevations they are active from June to July.

Larval biology: Larval burrows have been found concentrated in bare patches of soil in the mountain meadow habitat, but neither the larvae themselves nor their life history has been described.

White-striped Tiger Beetle, *Parvindela lemniscata* LeConte
(Plate 21) [Map 91]

Description and similar species: Length 7–9 mm; above bright metallic orange-red with a bold and straight yellowish-white stripe that runs the length of each elytron along the outer edge. The only other species in North America sharing this small size and a distinct, light yellowish stripe down each elytron is the closely related Grass-runner Tiger Beetle, which shares many of the same, upland desert grasslands, but is green in body color and rarely flies. The only other small tiger beetle in North America with a similar red color is the Sonoran Tiger Beetle, but it has greatly reduced maculations on the elytra with no long stripe. The ranges of these two species overlap broadly in southern Arizona and northern Mexico, but the Sonoran Tiger Beetle is more restricted to vegetated pond and stream edges and occurs less commonly in upland grassy areas.

Subspecies and morphological variants: Three very similar subspecies have been described, only two of which occur in the United States.

P. lemniscata lemniscata **LeConte:** This population occurs in the desert southwest of southeastern California, Arizona, and southwestern New Mexico south into western Mexico. Its legs are entirely yellowish-orange.

P. lemniscata rebaptisata **Vaurie:** This population occurs from southeastern New Mexico and western Texas south into central and northeastern Mexico. The femora of its legs are at least partially metallic red or green. A relatively broad zone of intergradation between these two subspecies occurs in southwestern New Mexico.

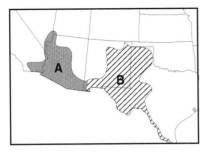

Map 91 White-striped Tiger Beetle, *Parvindela lemniscata*; **A**, *P. l. lemniscata*; **B**, *P. l. rebaptisata*.

Distribution and habitats: Restricted to open grasslands of the Sonora and Chihuahua Deserts, this species is often common to abundant in a variety of habitats. In addition to grassy upland areas away from water, it frequently occurs at the edge of rain puddles and ponds.

Behavior: Its small size and fast-running ability combine with the narrowed appearance of the body caused by the longitudinal yellow stripe on each elytron to make it easily mistaken for large red ants in the genus *Pogonomyrmex*, which sting painfully and are abundant in the same habitats. If in danger, individuals quickly fly short distances to escape. The small size of the White-striped Tiger Beetle also allows it to quickly thermoregulate in the hot desert sun, and thus it is active throughout most of the day, even when larger, less efficiently thermo-regulating tiger beetles are forced to seek shade or burrow into the substrate. This small tiger beetle frequently flies to lights at night. At times following a summer rain storm, it can be found by the hundreds around bright outdoor lights, such as at isolated gasoline filling stations.

Seasonality: Active from July to October the summer rains influence the height of local activity. Overwinters only as larvae. The life cycle is 1 or 2 years.

Larval biology: Burrows occur in open grassland on sandy or sandy-clay soils. Burrow depths are 12–20 cm.

Grass-runner Tiger Beetle, *Parvindela debilis* Bates
(Plate 21) [Map 92]

Description and similar species: Length 8–10 mm; above dull dark green with a conspicuous white line running down the length of the outer edge of each elytron. Below dark metallic green. Has a similar elytral pattern and frequently occupies the same habitat as the White-striped Tiger Beetle, but the running behavior of the Grass-runner Tiger Beetle and green, not bright copper upperparts, distinguish it.

Subspecies and morphological variants: In Mexico, many individuals are black or dark brown, but no subspecies are presently recognized.

Map 92 Grass-runner Tiger Beetle, *Parvindela debilis*.

Distribution and habitats: A species of moist plateau grasslands, it is found in short grass prairies, bunch grass pastures and sandy fields along the western Mexican border. It rarely occurs along temporary pond edges that may form in its upland habitat.

Behavior: Although capable of short, feeble flights, this species usually sprints between grass clumps so quickly that unless you are lucky your best view will be a shadow flitting out of the corner of your eye. Active following significant rain fall, it ventures out into open bare areas only when the soil is moist. As the soil dries, individuals become more and more restricted to the remaining moist areas near grass clumps (and occasionally under dry cow droppings). After a week or so without rain, adults of the Grass-runner Tiger Beetle are almost impossible to observe as they burrow into the bases of grass clumps and become inactive until the next rainfall.

Seasonality: Adults are active from July to October, and overwinters as larvae.

Larval biology: Burrows are found in the same habitat as that of the adults in bare open patches and at the base of grass clumps in upland areas. Burrows are 6–12 cm deep.

Swift Tiger Beetle, *Parvindela celeripes* LeConte (Plate 21) [Map 93]

Description and similar species: Length 8–9 mm; above brownish with maculations reduced to dots or absent except for a narrow white band along rear fourth to half of each elytron edge. Below coppery-green. Body "spindle"-shaped with elytra expanding toward the rear, especially in the female (Fig. 4.18B). Legs metallic copper to green. Below coppery-green. Extremely similar to the closely related Ant-like Tiger Beetle, which has more extensive maculations and the band along the outer edge of the elytra extending almost its whole length. The legs of the Ant-like Tiger Beetle are mainly orangish-yellow with little metallic sheen. The two species ranges overlap only along the lower Missouri River Valley and perhaps in northern Arkansas.

Subspecies and morphological variants: No distinct geographical forms have been described.

Map 93 Swift Tiger Beetle, *Parvindela celeripes*.

Distribution and habitats: Known from the southern Great Plains west of the Missouri River. This species occupies grasslands, prairie hillsides, clay hilltops, sand hills, prairie bluffs, railroad embankments, open forest paths, and grassy areas near streams. Recently an isolated colony was found in north central Arkansas.

Behavior: Extremely wary, this tiny species is apparently flightless and runs quickly like a large ant among the bases of grass and vegetation to elude enemies. It hides at the base of thick vegetation. Because of its secretive habits, this species is easily overlooked and may be more common than presently thought. Most easily seen when it emerges from vegetation to cross bare areas of soil.

Seasonality: Adults have been found from May to August but most commonly in July.

Larval biology: Larva unknown.

Ant-like Tiger Beetle, *Parvindela cursitans* LeConte
(Plate 21) [Map 94]

Description and similar species: Length 7–8 mm; above brown with maculations reduced to dots and lines except a thin complete or interrupted white band running along most of the outer edge of the elytra. Below coppery metallic green. Body "spindle"-shaped with elytra expanding toward the rear (Fig. 4.19), but less so than the similar Swift Tiger Beetle. Legs are yellow-orange with little metallic sheen. Extremely similar to the closely related Swift Tiger Beetle, which has less extensive maculations and the band along the outer edge of the elytra extending only along the rear fourth to half of each elytron. The legs of the Swift Tiger Beetle are distinctly metallic green to copper. The ranges of the two species overlap only along the lower Missouri River Valley and perhaps in northern Arkansas.

Subspecies and morphological variants: No distinct geographical populations have been described.

Distribution and habitats: Occurring locally from the Ohio and lower Mississippi River basins west and north to the Upper Missouri River

Map 94 Ant-like Tiger Beetle, *Parvindela cursitans.*

basin, this species is usually found in the vicinity of rivers and streams, salt flats, wet meadows and roadside ditches on open to slightly shaded ground. In some parts of the range it occurs regularly on the floor of open forests. The soil is moist or wet and consists of clay, loam or sand and often with sparse vegetation. The presently known range suggests that this species does not have a continuous distribution from south to north. It is unknown from most of Missouri, but this apparent disjunct pattern could be due to the difficulty of finding individuals of this elusive species.

Behavior: Generally runs but recent reports indicate it is capable of long flights. When it runs it does so quickly across open spaces in the vegetation looking much like an ant. It hides at the base of dense vegetation to escape danger and is often hard to observe or capture.

Seasonality: Adults have been found active from June to August but most commonly in late June and July.

Larval biology: Larvae are in moist clay soils around the base of plants.

Leaf Litter Tiger Beetles, Genus *Apterodela*

Formerly placed in the genus *Cylindera*, the large size, other unusual characters, and DNA analysis of this species place it as the only American representative in the *Apterodela*, a group with three other species found in similar habitats in eastern Asia.

One-spotted Tiger Beetle, *Apterodela unipunctata* Fabricius (Plate 24) [Map 95]

Description and similar species: Length 14–18 mm; above dull brown with numerous green, shallow pits across the elytra.Maculations reduced to a single small, white spot midway along the elytral edge. In profile the elytra are very flat. Below metallic violet-blue and green.

Subspecies and morphological variants: No distinct subspecies have been described.

Map 95 One-spotted Tiger Beetle, *Apterodela unipunctata*.

Distribution and habitats: This woodland species actively forages in and under the leaf litter of shaded forests from Missouri east to the mid-Atlantic coast and north up the Hudson River Valley of New York.

Behavior: This species usually runs across bare areas or stays motionless up to 10 minutes camouflaged on top of dead leaves or hiding under leaf litter. They are reported to take flight to escape from predators, but this ability may be restricted to certain geographic areas, including the Carolinas. Most active in late afternoon and early evening. Reported to be active at night. On cloudy overcast days occasionally ventures out into open bare patches of soil and roadways. Usually solitary and easily overlooked.

Seasonality: Adults active from April to September but mostly June and July. Overwinters as larvae and has a 2-year life cycle.

Larval biology: Burrows in steep, sparsely forested hillside with bare, rocky soil and in open patches of forest with sandy-clay soil. Many tunnels are vertical and then turn horizontal towards their bottoms. Burrow depth is 7–12 cm.

Dromo Tiger Beetles, Genus *Dromochorus*

All the species in this genus lack flight wings, but the oval elytra are not fused. They all are black or dark olive-brown, lack maculations and have only a few sparse and very small hair-like setae. The head is relatively large. They are unlike any other tiger beetles in the United States and Canada except the unrelated Night-stalking Tiger Beetles (*Omus*) of far western North America. They superficially resemble black species of Blister Beetles (Family Meloidae) or Darkling Beetles (Family Tenebrionidae), but are much swifter and more long-legged. The genus is restricted to south central North America. DNA differences often separate species of this genus that are difficult to tell apart externally. Thus microhabitats and geographical distributions become important distinguishing field characters.

Frosted Tiger Beetle, *Dromochorus pruininus* Casey
(Plate 23) [Map 96]

Description and similar species: Length 12–14 mm; above black and, in the hand, a smooth blue-violet velvet-like texture on the elytral surface lacks both granular microsculpturing and shallow green pits. Male labrum black with central portion ivory-colored (Fig. 4.8). Female labrum metallic black. Maxillary palps pale or yellow with contrasting dark terminal segment (Fig. 4.9A). The similar Loamy-ground Tiger Beetle has a distinctive granular or pitted texture to the elytral surface with no blue-violet tint, and the maxillary palps are typically darker.

Subspecies and morphological variants: No subspecies are recognized.

Map 96 Frosted Tiger Beetle, *Dromochorus pruininus*.

Distribution and habitats: Found from Kansas (one hypothetical record from Nebraska) and locally in Missouri to central Texas where it approaches but does not overlap with the range of the Loamy-ground Tiger Beetle. Further studies, such as microhabitat differences for larvae and adults, are needed to help better distinguish the distribution of the Loamy-ground Tiger Beetle and the Frosted Tiger Beetle. The Frosted Tiger Beetle occupies disturbed habitats, meadows, cultivated fields, lawns, parks, forest paths and clearings.

Behavior: Most active in early morning and late afternoon to evening but can be found active throughout the day. Flightless, it runs swiftly through dense vegetation and is evident only when it comes out onto cleared patches of soils or roadways. Occasionally hides in cracks in the ground or burrows under flat rocks to escape danger.

Seasonality: Adults active from May to August and overwintering is in the larval stage. It probably has a 2-year life cycle.

Larval biology: Larvae have been raised in captivity, but the location of larval burrows and their microhabitat in the wild remain unknown.

Loamy-ground Tiger Beetle, *Dromochorus belfragei* Sallé
(Plate 23) [Map 97]

Description and similar species: Length 12–14 mm; above black and, in the hand, a granular or pitted texture on the elytral surface; southern and eastern populations have sutural punctures. Male labrum black with central portion ivory-colored (Fig. 4.8). Female labrum metallic black. Maxillary palps dark yellow to reddish-brown with contrasting dark terminal segment (Fig. 4.9A). The similar Frosted Tiger Beetle has a smoother and more velvet-like elytral surface with a violet-blue tint. Under magnification, the lack of granular texture on the elytra of the Frosted Tiger Beetle is distinctive. These two species apparently do not overlap but come close in parts of Oklahoma and northern Texas. Many more observations are needed to determine their ranges precisely. The extremely similar Juniper Grove Tiger Beetle has metallic green, gold, or blue reflections in a row of sutural punctures, but is found only in the Hill Country of central Texas in upland juniper grove habitat.

Subspecies and morphological variants: There is no evidence of intermediate forms between the Frosted and the Loamy-ground Tiger Beetles, even though at some sites, such as in the Dallas area of Texas and south central Oklahoma, both forms have been collected within a few kilometers of each other. DNA data reinforce the interpretation that these two species are genetically distinct and appear not to hybridize. In the southern part of the Loamy-ground Tiger Beetle's range, variation in the pattern of granular texture on the elytral surface may indicate a subspecies or perhaps another cryptic species is present.

Map 97 Loamy-ground Tiger Beetle, *Dromochorus belfragei.*

Distribution and habitats: Although the distinctive granular elytral surface and jet black color of the Loamy-ground Tiger Beetle readily distinguish it from the smoother elytral surface and blue-violet reflections of the Frosted Tiger Beetle, subtle differences in microhabitat use and behavior remain unclear. Both species appear to share a habitat of grasslands, sodded fields, hilltops, slopes, vacant lots, roadside ditches, river and stream banks. Adults move from upland areas, such as hilltops, to lower areas closer to water in mid-summer.

Behavior: Active throughout the day, especially on overcast days, but primarily in the afternoon until dusk. Flightless, it runs swiftly through dense vegetation and is evident only when it comes out onto cleared patches of soils or roadways. Easily overlooked.

Seasonality: Adults active from May to August and overwintering is in the larval stage.

Larval biology: Larva unknown.

Juniper Grove Tiger Beetle, *Dromochorus knisleyi* Duran et al. (Plate 23) [Map 98]

Description and similar species: Length 12–14 mm; above black and, in the hand, a granular or pitted texture on the elytral surface with metallic gold, green, or blue reflections in the sutural punctures. Most individuals have textured "swirls" (infuscations) on the central part of the elytra (Fig. 4.7). Male labrum black with central portion ivory-colored. Female labrum metallic black. The extremely similar Loamy-ground Tiger Beetle has no metallic colored reflections in sutural pits, has paler maxillary palps (except the terminal segment), and is not found in the Hill County region of Texas.

Subspecies and morphological variants:No distinct geographical populations have been described.

Map 98 Juniper Grove Tiger Beetle, *Dromochorus knisleyi*.

Distribution and habitats:Found only in the Hill Country region of central Texas. It is restricted to shaded juniper groves in upland hilly areas.

Behavior:Little is known about the biology of this recently described species. It has been found active from afternoon until dusk in shaded and semishaded grassy areas in and around juniper groves on high dry hillsides. It appears restricted to grassy patches underneath juniper trees and does not venture far from the shade of the branches.

Seasonality: Active from May–June but additional data are needed to better characterize the seasonality of this species.

Larval biology:Larva unknown.

Gulf Prairie Tiger Beetle, *Dromochorus welderensis Duran et al.* (Plate 23) [Map 99]

Description and similar species:Length 12–15 mm; above black and, in the hand, a smooth velvet-like texture on the elytral surface that lacks sutural punctures. Upper surface jet black, rarely with faint blue reflections. Male labrum black with central portion ivory-colored (Fig. 4.8). Female labrum metallic black. Maxillary palps dark brown to black throughout, sometimes with metallic green and violet reflections (Fig. 4.9B). The extremely similar Chaparral Tiger Beetle is nearly indistinguishable in appearance, but occurs more inland in chaparral scrub habitat. The Velvet Tiger Beetle has a narrower body, a strong violet-blue sheen along the edges of the elytra, and an all dark labrum in the male. The Loamy-ground Tiger Beetle has a rougher granular texture, has paler maxillary palps (except the terminal segment) (Fig. 4.9A), and is found more inland. None of these similar species is known to overlap in range with the Gulf Prairie Tiger Beetle or occur in the same habitat.

Subspecies and morphological variants: No distinct geographical populations have been described.

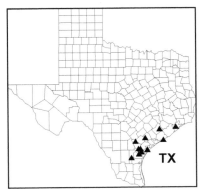

Distribution and habitats: Found in Gulf prairie habitat of dense clumps of grasses interspersed with bare patches of dark soil; coastal Texas from Houston to Corpus Christi.

Behavior: Flightless, this species often hides in clumps of dense grass and may come out to forage in adjacent patches of bare ground in the early morning or early evening or throughout the day when overcast.

Map 99 Gulf Prairie Tiger Beetle. *Dromochorus welderensis.*

Seasonality: Active from May to August, with peak activity in June.

Larval biology: Larva unknown.

Chaparral Tiger Beetle, *Dromochorus chaparrelensis* Duran et al. (Plate 23) [Map 100]

Description and similar species: Length 12–15 mm; above black and, in the hand, a smooth velvet-like texture on the elytral surface that lacks sutural punctures. Upper surface jet black, rarely with faint blue reflections. Male labrum black with central portion ivory-colored (Fig. 4.8). Female labrum metallic black. Maxillary palps dark brown to black throughout (Fig. 4.9B), sometimes with metallic green and violet reflections. The extremely similar Gulf Prairie Tiger Beetle is nearly indistinguishable in appearance, but occurs in coastal Gulf prairie habitat. The Velvet Tiger Beetle has a narrower body, a strong violet-blue sheen along the edges of the elytra, and an all dark labrum in the male.

Indistinguishable in appearance from the Pygmy Dromo Tiger Beetle, except for average size. These two species occur in grasses and bare patches around mesquite trees and scrub vegetation, but apparently do not overlap in range. Molecular data indicate that they are genetically distinct. Little is known about either of these species, and more observations on behavior and microhabitat are needed to better understand how their ecological roles might differ.

Subspecies and morphological variants: No distinct populations are recognized.

Map 100 Chaparral Tiger Beetle, *Dromochorus chaparrelensis*.

Distribution and habitats: Presently known to occur only in mesquite chaparral savanna from Dimmit County, Texas to Nuevo Laredo, Mexico. The range of this species is likely to be larger, especially in Mexico, but more data are needed.

Behavior: Very rarely observed and poorly known. Occurs in bare patches of ground surrounded by grasses and scrub vegetation.

Seasonality: Active from April to July. Peak activity apparently in late May to early June.

Larval biology: Larva unknown.

Pygmy Dromo Tiger Beetle, *Dromochorus minimus* Duran et al. (Plate 23) [Map 101]

Description and similar species: Length 10–13 mm; above ash-colored to black and elytral surface smooth and velvet-like in texture with no sutural punctures. Upper surfaces with an ashy-gray, beige, or bluish tint. Male labrum black with central portion ivory-colored (Fig. 4.8). Female labrum metallic black. Maxillary palps dark brown to black throughout (Fig. 4.9B), sometimes with metallic green and violet reflections. Often smaller than other species of *Dromochorus*, but otherwise extremely similar to the Chaparral Tiger Beetle (see above species account). The similar Frosted Dromo Tiger Beetle has yellow or pale maxillary palps with a contrasting dark terminal segment (Fig. 4.9A), and its body is often larger.

Subspecies and morphological variants:No distinct geographical populations have been described.

Distribution and habitats: Known only from mesquite chaparral savanna habitat in Bexar, Frio, and Atascosa counties of south Texas. The range of this species may be larger, but more data are needed.

Behavior: Rarely seen and poorly known. It is an incredibly fast runner, even for a Dromo Tiger Beetle. It appears to prefer grassy areas and bare patches under the shade of mesquite trees.

Map 101 Pygmy Dromo Tiger Beetle, *Dromochorus minimus.*

Seasonality: Active from May to June. Apparently most common in late May.

Larval biology: Larva unknown.

Velvet Tiger Beetle, *Dromochorus velutinigrens* W. N. Johnson (Plate 23) [Map 102]

Description and similar species: Length 12–15 mm; body is noticeably more slender than other Dromo Tiger Beetles; above black and elytral surface smooth and velvet-like in texture with no small pits. Elytral surface is dark violet-blue along the edges and rear tip. Male labrum completely metallic black with a green sheen. Female labrum metallic black. Maxillary palps dark brown to black throughout, sometimes with metallic green and violet reflections (Fig. 4.9B). The similar Gulf Prairie Tiger Beetle and Chaparral Tiger Beetle are not as slender and are usually all black without violet-blue reflections on the edges of the elytra. Males of these latter species have a labrum with a central portion ivory-colored (Fig. 4.8).

Map 102 Velvet Tiger Beetle, *Dromochorus velutinigrens.*

Subspecies and morphological variants: No distinct geographical populations have been described.

Distribution and habitats: Found in south Texas along the coast. Likely occurs in adjacent Mexico as well, but there are no records south of the Rio Grande as of yet. Occurs along sandy roads and paths in grassy regions in open forest and coastal savanna, as well as coastal salt flats, salt marshes, gulf prairies and clay dunes.

Behavior: Flightless, crepuscular and nocturnal, it runs swiftly through dense vegetation and is evident only when it comes out onto cleared patches of soils or roadways.

Seasonality: Active from April to mid-June following measurable rainfall.

Larval biology: Larva unknown.

Cajun Tiger Beetle, *Dromochorus pilatei* Guérin-Méneville (Plate 23) [Map 103]

Description and similar species: Length 12–14 mm; above dark olive-brown to black and, in the hand, a granular texture on the elytral surface interspersed with distinct, sutural punctures reflecting bright green. Some individuals have additional extensive green reflections on the upper surfaces, especially on or near the head. Male labrum completely ivory-colored sometimes with a black border. Female labrum metallic black.

Subspecies and morphological variants: No distinct geographical populations have been described.

Distribution and habitats: Known from coastal northeastern Texas (Galveston County) to coastal and central Louisiana (Natchitoches Parish). Here it is found along forested paths, fields, under vegetative detritus and agricultural fields usually near water, such as bayous, lakes and drying salt marshes.

Behavior: Flightless, primarily crepuscular, it runs swiftly through dense vegetation and is evident only when it comes out onto cleared patches of soils

Map 103 Cajun Tiger Beetle, *Dromochorus pilatei.*

or roadways. May be active throughout the day when overcast or when found in shadier microhabitats. Easily overlooked.

Seasonality: Adults active from May to July and overwintering is in the larval stage.

Larval biology: Larva biology unknown.

Ellipsed-winged Tiger Beetles, Genus *Ellipsoptera*

The thirteen species of this genus all occur in the United States and/or Canada, with ranges of six species extending into northern Mexico or Cuba. They share characteristics, such as DNA, extremely protruding eyes, long legs and a peculiarly long and circular flagellum in the male genitalia.

Coastal Tiger Beetle, *Ellipsoptera hamata* Audouin and Brullé (Plate 24) [Map 104]

Description and similar species: Length 9–13 mm; above shiny olive to rusty-brown with three distinct maculations that are joined by a wide line along the outer edge of the elytra. The middle maculation is distinctively diffuse, a character shared only with the similar and closely related Margined Tiger Beetle. The two species occur together on the west coast of peninsular Florida. In the hand, they can be separated by the lack of a peculiar tooth-like extension on underside of the right mandible of male Coastal Tiger Beetles (Fig. 4.11B) and the lack of a distinctive right angle downturn to the tip of the elytra on female Coastal Tiger Beetles (Fig. 4.11D).

Subspecies and morphological variants: Four subspecies have been described, but two are restricted to Mexico.

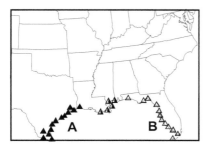

Map 104 Coastal Tiger Beetle, *Ellipsoptera hamata*; **A**, *E. h. monti*; **B**, *E. h. lacerata*.

E. hamata lacerata Chaudoir: Rusty-brown above and occurs from the Florida Keys north and west to Alabama. Intergradation with *E. h. monti* from Alabama to Louisiana.

E. hamata monti Vaurie: Olive above and occurs west and south to Texas and into northeastern Mexico.

Distribution and habitats: A Gulf Coast species limited to tidal marshes, coastal mudflats, bare or sparsely vegetated mud near the water's edge, coarse sand beaches and creek shores emptying into the ocean, this species has not been recorded inland, except in extreme southern Texas.

Behavior: Often gregarious, it associates with the Margined Saltmarsh Tiger Beetle, Gulfshore Tiger Beetle, and S-banded Tiger Beetle. It escapes danger primarily by running and then flying short distances, often looping out over the water before circling back. Active both during the day and night, it is commonly attracted to lights at night. Regularly scavenges dead fish and other organisms washed up on the beach.

Seasonality: Adults have been found active from April to December but most commonly in June. The species overwinters as larvae.

Larval biology: Larva unknown.

Margined Tiger Beetle, *Ellipsoptera marginata* Fabricius
(Plate 24) [Map 105]

Description and similar species: Length 10–13 mm; above green to olive-brown and rarely black with three maculations that are joined along the edge of the elytra. The middle maculation is distinctively diffuse. In the hand, the rear tip of the elytra in the females is peculiarly turned down almost ninety degrees (Fig. 4.11C), and the male has an equally peculiar tooth-like extension on the bottom of the right mandible (Fig. 4.11A). Metallic copper-green below with the sides of the thorax, abdomen, and cheeks on the head covered with dense, white, hair-like setae.

Subspecies and morphological variants: No distinct subspecies have been described.

Distribution and habitats: Restricted to the sea coast from New England to the panhandle of Florida. It occupies habitats such as coastal mudflats, sandy ocean beaches, shores of salt marshes and mouth of streams emptying into the ocean. In New England, apparently it has been extirpated from all but a few protected areas. A specimen collected recently in Nova Scotia may have been a storm-blown individual. Regularly found together with the Eastern Beach Tiger Beetle and the Hairy-necked Tiger Beetle.

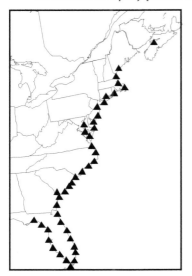

Map 105 Margined Tiger Beetle, *Ellipsoptera marginata.*

Behavior: Very wary and flies up to avoid danger at a distance of 5 m or more. Often flies out over water and lands in shallow water to swim back to shore. Active both day and night and commonly attracted to lights at night.

Seasonality: Adults active from February (in far south) to September but most common in June and July.

Larval biology: Relatively shallow (8–15 cm) burrows are found in sandy sea beach a few meters back from high tide line in open, sparsely vegetated soil.

Sandbar Tiger Beetle, *Ellipsoptera blanda* Dejean
(Plate 24) [Map 106]

Description and similar species: Length 11–13 mm; above green-bronze with diffused maculations expanded and coalesced to cover most of the elytral surface. Below coppery with dense, white, hair-like setae covering the sides of the abdomen, thorax, and cheeks on the head. Similar to the Coastal Tiger Beetle and Margined Tiger Beetle, but the expanded maculations and inland range of the Sandbar Tiger Beetle distinguish it. Overlaps and shares the same habitat in the Gulf States with White-sand Tiger Beetle, which has the expanded white maculations as a broad band along the edge of the elytra and a solid brown stripe down the length of the middle of the elytra.

Subspecies and morphological variants: No subspecies of this range-restricted species are known.

Distribution and habitats: Found only along sandy beaches and white sand bars of black water rivers, and occasionally clay or mud banks and clearings above these rivers during flooding in the coastal plain of the southeastern United States.

Map 106 Sandbar Tiger Beetle, *Ellipsoptera blanda*.

Behavior: Often gregarious and present in large numbers. Active even on warm rainy days. Attracted to lights at night.

Seasonality: Adults active from May to September but mainly June and July. Apparently overwinters as larvae.

Larval biology: Larva unknown.

White-sand Tiger Beetle, *Ellipsoptera wapleri LeConte*
(Plate 24) [Map 107]

Description and similar species: Length 9–11 mm; above dark brown with the maculations coalesced into a broad white band along the edge of the elytra. Below metallic green with dense, white, hair-like setae covering the sides of the abdomen, thorax, and cheeks of the head. Overlaps and shares the same habitat in the Gulf States with Sandbar Tiger Beetle, which has expanded white maculations diffusely covering most of the elytral surface. Amazingly similar to eastern populations of White-cloaked Tiger Beetle, the two species are not likely to occur in the same habitat or overlap in range, but the White-sand Tiger Beetle tends to have thin white lines extend into the dark central area of the elytra and the White-cloaked Tiger Beetle does not. This dark area on the elytra is greenish-brown in the White-cloaked Tiger Beetle and reddish-brown in the White-sand Tiger Beetle.

Subspecies and morphological variants: No distinct subspecies have been described for this range-restricted species.

Map 107 White-sand Tiger Beetle, *Ellipsoptera wapleri.*

Distribution and habitats: Endemic to the coastal plains of the eastern Gulf of Mexico, it occurs on white sandy beaches of small and moderate-sized streams usually close to the water's edge. More common and widespread within its range than the Sandbar Tiger Beetle, with which it regularly occurs where their ranges overlap.

Behavior: Usually present in small numbers and not gregarious. Its escape flights are relatively short. Most common close to the water's edge, especially during the heat of the day.

Seasonality: Adults active from May to October but most common in June and July.

Larval biology: Larva unknown.

Nevada Tiger Beetle, *Ellipsoptera nevadica* LeConte
(Plate 24) [Map 108]

Description and similar species: Length 10–13 mm; above shiny dark brown, brown, reddish-brown or rarely dark green or blue. Three distinct maculations are connected by a white line along the edge of the elytra. Below metallic copper and green. Elytra distinctly widened, especially in the females, toward the middle. Diagnostic for all populations of this species is the top or front of the front maculation. It is absent and thus formed into the shape of a "J" and not a "G," which is the shape of this maculation in other similar species, such as the Sandy Stream Tiger Beetle and the Coppery Tiger Beetle. This "J"-shaped front maculation is also shared with the Rio Grande Tiger Beetle and the Aridland Tiger Beetle, two species with which the Nevada Tiger Beetle regularly occurs. Both the Rio Grande Tiger Beetle and the Aridland Tiger Beetle, however, have longer and straighter elytra that are parallel-sided, and their color above is dull brown, green or red and not so shiny as in the Nevada Tiger Beetle. The underside of the Aridland Tiger Beetle is bright metallic green or blue. In the hand and under magnification, the first antennal segment of the Nevada Tiger Beetle has many small setae and only one or a few long sensory setae (Fig. 4.13A).

Subspecies and morphological variants: Eight subspecies are presently recognized, and several additional distinct populations will probably be named with further studies. One subspecies, *E. n. metallica* **Sumlin**, is restricted to Coahuila, Mexico. The dark color above and the width of the maculations are the main characters used to distinguish populations. When DNA, elytral surface patterns and genitalia are finally analyzed, this presently recognized species may prove to be a complex of three or four cryptic species.

E. nevadica nevadica **LeConte:** Above coppery-bronze with green reflections with wide middle maculation connected, if at all, by a thin white line along the elytral edge to the other maculations. This form occurs patchily on isolated salt lakes from southeastern Nevada through the interior deserts of eastern California to northwestern Sonora, Mexico. In northern Nevada (Humboldt County) the maculations are so reduced that the first one is sometimes only a short line along the edge of the elytra and the other two are not connected by any line along the edge of the elytra.

E. nevadica citata **Rumpp.** Above bronzy-brown with green reflections, and maculations thin. This isolated population is restricted to southeastern Arizona and adjacent New Mexico.

E. nevadica knausii **Leng.** Above reddish-brown to brown with moderate maculations. The base of the middle maculation is connected along the edge of the elytra broadly forward to the front maculation but very narrowly backward to the rear maculation. This subspecies is the most widely ranging form, occurring the length of the Great Plains from just across the Canadian border in the Prairie Provinces south to central Texas. In northeastern New Mexico (Colfax County), many individuals are dark green above.

E. nevadica lincolniana **Casey:** Restricted to several inland salt marshes near Lincoln, Nebraska, this subspecies is characterized by being dark greenish-brown above and having reduced maculations that are almost absent in some individuals. **NOTE:** This subspecies is federally listed in the United States as ENDANGERED, and it is illegal to collect.

E. nevadica makosika **Spomer:** A distinct but extremely restricted population recently described from southwestern South Dakota (Pennington County) has significantly broadened maculations.

E. nevadica olmosa **Vaurie:** Above greenish-brown to brown and maculations very wide. Apparently disjunct range from coastal south Texas northwest along the Rio Grande to southern New Mexico. The population in southern New Mexico may be considered a separate subspecies, but these two populations may be connected through northern Mexico. This subspecies is separated by the Edwards Plateau in central Texas from *E. n. knausii* in northern Texas, and no intergrades are known.

E. nevadica tubensis **Cazier:** Above reddish with often expanded maculations. This subspecies occurs in the high desert and grasslands area of the eastern section of the Great Basin in the four-corners area of Utah, Colorado, New Mexico and Arizona.

E. nevadica **subspecies A:** The populations occurring on a few white salt lakes in central New Mexico (Torrance County) and northwestern Texas typically have greatly expanded maculations that diffusely cover much or most of the elytral surface in many individuals. This character of greatly expanded maculations is shared at these salt-covered lakes by several other unique populations of species that have been named as separate subspecies, such as the *estancia* form of Williston's Tiger Beetle and the *fascinans* form of the White-cloaked Tiger Beetle.

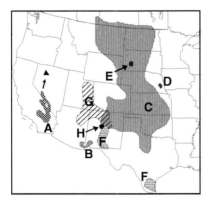

Map 108 Nevada Tiger Beetle, *Ellipsoptera nevadica*; **A**, *E. n. nevadica*; **B**, *E. n. citata*; **C**, *E. n. knausii*; **D**, *E. n. lincolniana*; **E**, *E. n. makosika*; **F**, *E. n. olmosa*; **G**, *E. n. tubensis*; **H**, *E. n.* ssp.

Distribution and habitats: Usually found on wet soil that is alkali or salty with sparse vegetation. Occurs along ponds, lakes, rivers, ditches and small to large salty spots in pastures and fields throughout western North America.

Behavior: Escapes danger with short flights. Solitary and active both day and night over most of its range but gregarious in the Great Plains. Regularly attracted to lights at night.

Seasonality: Adults active from April to August in the desert interior of

California, June to August in the Great Plains, July to September in the desert southwest, and June to November in the lower Rio Grande Valley of south Texas.

Larval biology: Burrows among vegetation near grass hummocks, margins of sloping banks, open flats of sandy or sandy-clay soils, and salt flats. Burrow depth 22–35 cm.

Coppery Tiger Beetle, *Ellipsoptera cuprascens* LeConte (Plate 25) [Map 109]

Description and similar species: Length 10–14 mm; above shiny copper-red to greenish-red. Maculations complete with "G"-shaped front maculations and a wide white line along the outer edge of the elytra that connects all three maculations. Below metallic copper-green. The similar Puritan Tiger Beetle is dull brown above and occurs only in the far eastern states and does not overlap with the Coppery Tiger Beetle. The extremely similar Sandy Stream Tiger Beetle is often greener and duller above where they overlap in the Midwest, but the reddish populations in the Great Plains can generally be separated only by careful examination of the elytral surface, which is not as shiny as that of the Coppery Tiger Beetle. In the hand, female Sandy Stream Tiger Beetles have the rear tips of the elytra pointed, with no notch between them. The rear tips of the elytra of female Coppery Tiger Beetles are rounded and have a notch. In addition, most individuals of the Coppery Tiger Beetle have the rear end of the middle maculation globular or only slightly enlarged. In the Sandy Stream Tiger Beetle, it is typically recurved or triangular. All populations of the Nevada Tiger Beetle and the Rio Grande Tiger Beetle lack the upper end of the front maculation and have it shaped instead in the form of a "J". In the hand and under magnification, the first antennal segment of the Nevada Tiger Beetle has many small setae and only one or a few long sensory setae (Fig. 4.13A).

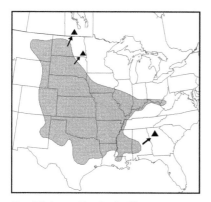

Subspecies and morphological variants: Although individuals with shiny reddish upperparts are more common in the western parts of its range and shiny greenish-red upperparts predominate in the east, there is no distinct separation and no basis upon which to distinguish subspecies.

Distribution and habitats: Closely associated with the edge of water on

Map 109 Coppery Tiger Beetle, *Ellipsoptera cuprascens.*

sandy beaches of rivers and streams. Occasionally on loam or mud flats away from the river's edge.

Behavior: Small numbers occur together but not normally gregarious. Readily attracted to lights at night.

Seasonality: Adults active from May to October but mainly in July.

Larval biology: Burrows have ragged, crater-like openings in sandy soils.

Sandy Stream Tiger Beetle, *Ellipsoptera macra* LeConte (Plate 25) [Map 110]

Description and similar species: Length 11–16 mm; above dull olive-green, reddish-brown, or reddish. Maculations complete with "G"-shaped front maculations and a wide white line along the elytral edge connecting all three maculations. The extremely similar Coppery Tiger Beetle is much shinier red above, and the rear end of the middle maculation is globular or not enlarged. In the Sandy Stream Tiger Beetle this maculation ends in a triangular or recurved point. In the hand, female Sandy Stream Tiger Beetles have the rear tips of the elytra pointed, with a no notch between them. The rounded rear tips of the elytra of female Coppery Tiger Beetle have a notch. All populations of the Nevada Tiger Beetle and the Rio Grande Tiger Beetle lack the upper end of the front maculation and have it shaped instead in the form of a "J".

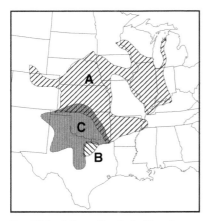

Map 110 Sandy Stream Tiger Beetle, *Ellipsoptera macra*; **A**, *E. m. macra*; **B**, *E. m. ampliata*; **C**, *E. m. fluviatilis*.

Subspecies and morphological variants: Three subspecies have been described based largely on the dark color above and the width of the maculations. Larger individuals occur in the southern parts of the species range.

E. macra macra LeConte: Above dark olive to reddish-green and brown with moderately wide maculations. The middle one is often eroded or diffuse. This form occurs north and east of Oklahoma throughout the Midwest. A zone of intergradation with *E. m. fluviatilis* runs along the length of southern Kansas and northern Arkansas.

E. macra ampliata Vaurie: Above dark olive-green with thin maculations. It is restricted to northeastern Texas north of Dallas.

E. macra fluviatilis Vaurie: Above red to reddish-brown with wide maculations. This subspecies is present in Oklahoma and most of northern Texas, except the northeastern part of the state.

Distribution and habitats: Open ground, mud and sand on beaches of rivers and large lakes such as Lake Michigan.

Behavior: Common and gregarious. Rests occasionally on short vegetation above the ground. Often attracted to lights at night.

Seasonality: Adults active from May to October but mostly in July.

Larval biology: Shallow burrows have smooth openings. They are located in sandy, sparsely vegetated areas above the river shoreline.

Puritan Tiger Beetle, *Ellipsoptera puritana* G. H. Horn
(Plate 25) [Map 111]

Description and similar species: Length 12–14 mm; above greenish-bronze with maculations connected along the elytral edge by a broad white line. The inner end of the middle band is usually eroded or ragged. Below coppery. Closely related and similar to the Sandy Stream Tiger Beetle, most individuals of the Puritan Tiger Beetle are browner above, but there is also no overlap in distribution. The nearest record of Sandy Stream Tiger Beetle is in southwestern Michigan, more than 800 km to the west. **NOTE:** This species is federally listed in the United States as THREATENED, and it is illegal to collect.

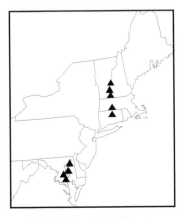

Map 111 Puritan Tiger Beetle, *Ellipsoptera puritana.*

Subspecies and morphological variants: Although there are no reliable external characters to separate northern and southern populations, molecular analysis shows that the New England populations are so genetically distinct from the populations in the Chesapeake that they may actually represent two separate species.

Distribution and habitats: Formerly occurred along narrow beaches of rivers of the Connecticut River from New Hampshire to Connecticut. It disappeared from New Hampshire in the late 1920s and most other New England sites by the

1950s. Along the Connecticut River now restricted to two disjunct sites in southern Massachusetts and Connecticut. Also found on Maryland beaches of the Chesapeake Bay below the Calvert Cliffs (Calvert County) and near the mouth of the Sassafras River (Kent and Cecil counties).

Behavior: Often gregarious and apparently a good disperser as individuals have been found up to 45 km from established colonies. Rarely found away from water's edge. Attracted to lights at night.

Seasonality: Adults active from June to September but most abundant from late June to early August.

Larval biology: The larval habitat is different in the two disjunct locations. In New England they are found close to the water's edge on sandy river flood plains, but in Maryland they are restricted to vertical faces of the upper cliff layers in sandy soils.

Rio Grande Tiger Beetle, *Ellipsoptera sperata* LeConte
(Plate 25) [Map 112]

Description and similar species: Length 11–13 mm; above dull brown to dark brown. Maculations moderate to thin and connected by a line along the edge of the elytra. The front maculation lacks a top or front section and is shaped like a "J". The body shape is long with straight and parallel-sided elytra in the males with a slight rounding in the females. Below dark coppery. Two other similar species share the "J"-shaped front maculation. The Nevada Tiger Beetle above is shiny and the body shape is more distinctly expanded at the mid elytra. Under magnification, the Nevada Tiger Beetle also has distinct setae on the first antennal segment. The Aridland Tiger Beetle is bright green or red above and metallic green or blue below.

Subspecies and morphological variants: Three subspecies have been described, but one of these, *E. s. vauriei* Cazier is restricted to Sonora, Mexico.

    E. sperata sperata LeConte: Above dull brown with moderate width maculations. The rear part of the elytra comes to a very narrow point in the female. This subspecies occurs along the Rio Grande from south Texas to New Mexico, and then into northeastern Arizona and eastern Utah. Remnant populations occur along a few moist stretches of the Salt River and lower Colorado River in southern Arizona and extreme southeastern California.

    E. sperata inquisitor Casey: Above dull dark brown with thin maculations. The rear part of the elytra comes to a broad point in the female. It is found in central Texas and there is little or no intergradation with nominate forms along the southern coast of Texas. Some experts consider this population to be a distinct species.

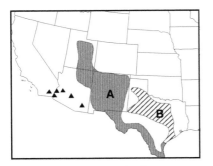

Map 112 Rio Grande Tiger Beetle, *Ellipsoptera sperata*; **A**, *E. s. sperata*; **B**, *E. s. inquisitor*.

Distribution and habitats: Muddy and sandy banks and beaches of rivers, salt marshes, meadows and paths close to streams. Usually bare soil or sparsely vegetated areas.

Behavior: Often very gregarious and attracted to lights at night.

Seasonality: Adults active June to August in central Texas and May to September elsewhere.

Larval biology: Larva unknown.

Aridland Tiger Beetle, *Ellipsoptera marutha* Dow
(Plate 25) [Map 113]

Description and similar species: Length 11–13 mm; above bright green or rusty-red with thin but complete maculations connected by a white line along the edge of the elytra. The front maculation lacks a top or front section and is shaped like the letter "J". The body shape is long with straight and parallel-sided elytra in the males with a slight rounding in the females. Below metallic green or blue. Two other similar species share the "J"-shaped front maculation. The Nevada Tiger Beetle is shiny above and the body shape is more distinctly expanded at the mid elytra. The Rio Grande Tiger Beetle is dull brown above and coppery below.

Map 113 Aridland Tiger Beetle, *Ellipsoptera marutha*.

Subspecies and morphological variants: Although no subspecies have been distinguished, two very different color forms occur. Bright green individuals are most common in the southern parts of its range, and bright rusty-red individuals are more common in the northern parts of its range, but both forms occur together in many areas.

Distribution and habitats: Occurs on the edges of temporary and permanent ponds, reservoirs, alkaline lakes, shallow grassland

193

streams, drainage ditches, watering tanks, salt flats, irrigated fields, and marshy grasslands. Adults often appear far from water on sandy ridges and dunes where they oviposit, usually at night.

Behavior: Highly gregarious during the day. Commonly attracted to lights at night and a long distance disperser at night.

Seasonality: Adults active from June to September but mostly in July at the beginning of the summer rains. Overwinter as larvae and has a 2- to 3-year life cycle.

Larval biology: Larval burrows up to 1 km away from water's edge in upland sand dunes surrounded by grasslands. Burrows are deep (18–45 cm) with moist soil usually at the bottom. Larvae are active primarily at night and plug the entrances to their burrows during hot, dry days. Larvae are attacked both by bee-fly (*Anthrax*) (Fig. 7.2) and tiphiid wasp (*Pteromborus*) parasitoids.

Ghost Tiger Beetle, *Ellipsoptera lepida* Dejean
(Plate 25) [Map 114]

Description and similar species: Length 9–11 mm; above the metallic green, bluish or coppery dark colors are limited to head and thorax. Maculations of elytra expanded and coalesced to diffusely cover the entire elytral surface. Below greenish-bronze and covered with dense, white, hair-like setae. Whitish legs diagnostic. Sides of elytra distinctly expanded.

Subspecies and morphological variants: No subspecies are currently recognized for this extremely widespread species. Most populations have individuals of all three background colors, coppery, blue or green, but the proportion of individuals exhibiting each of these colors shows no distinct geographical pattern.

Distribution and habitats: Distributed patchily from New York to Nevada but likely absent from the southeastern states. It occurs only in deep, loose sand fields such as coastal and inland dunes, sandy washes, sand ridges through open forest, and well-drained, dry, sparsely vegetated soil. Now absent from many of its historical breeding sites, probably due to sand excavation, land development and stabilization of sand dunes by encroaching vegetation.

Map 114 Ghost Tiger Beetle, *Ellipsoptera lepida*.

Behavior: Usually solitary and present at low densities. Dependent on extreme camouflage, it often "freezes" in position at the approach of danger depending on its not being seen by the potential predator. Usually its flights are short (2–5 m), but occasionally it flies vertically to be caught by the wind and carried great distances. Regularly attracted to lights at night and has been found actively feeding in the evening as well as during the daytime. Found on dune crests, slopes and interdunal bowls. During heat of day adults dig shallow burrows into sand beneath tufts of grass.

Seasonality: Adults active from March to October but most common in June and July. Overwinters as larvae and has a 2- or 3-year life cycle.

Larval biology: Burrows found in sheltered bowl areas of sand dunes or bank sides with drifting sand. Burrows may be very deep (0.5–3 m).

Whitish Tiger Beetle, *Ellipsoptera gratiosa* Guérin-Méneville (Plate 25) [Map 115]

Description and similar species: Length 10–12 mm; above bronze but top of head and thorax so densely covered with white, hair-like setae the dark color underneath is difficult to see. The bronze color on the elytra is restricted to a narrow band down the length of the middle. The maculations are so expanded that they cover most of the elytral surface in white. Labrum with few or no setae. Bronzy below but densely covered with white setae. Closely resembles the related Moustached Tiger Beetle, which is smaller, has more ragged edges where the inner border of the white meets the bronze of the central elytra, and has its labrum completely covered with white setae. The ranges of these two species overlap only in southeastern Georgia and northern Florida. Another species, the Ghost Tiger Beetle, has almost all-white elytra and lacks a distinctive central dark stripe down the length of the middle of the elytra. However, it does not occur in the southeastern United States.

Subspecies and morphological variants: No subspecies have been described.

Distribution and habitats: Occurs in dry, open, sandy areas on paths, roads and forest clearings in pine barrens on the coastal plains of the central and southern Atlantic and eastern Gulf of Mexico.

Map 115 Whitish Tiger Beetle, *Ellipsoptera gratiosa*.

Behavior: Fast runner and flies only short distances to escape danger.

Seasonality: Adults active from March to October but mostly June to September.

Larval biology: Vertical burrows dug in dry sand mixed with clay to sandy-clay soils with sparse vegetation. Burrow depth 55–110 cm.

Moustached Tiger Beetle, *Ellipsoptera hirtilabris* LeConte (Plate 25) [Map 116]

Description and similar species: Length 9–11 mm; above bronze but top of head and thorax so densely covered with white, hair-like setae the dark color underneath is difficult to see. The bronze color on the elytra is restricted to a narrow band down the length of the middle. The maculations are so expanded that they cover most of the elytral surface in white. Labrum densely covered with white setae. Bronzy below but densely covered with white setae. Closely resembles the related Whitish Tiger Beetle, which is larger and has sharper edges where the inner border of the white meets the bronze of the central elytra. Diagnostic in the hand, however, the labrum of the Whitish Tiger Beetle has few or no white hairs. The range of the Moustached Tiger Beetle overlaps with that of the Whitish Tiger Beetle only in southeastern Georgia and northern Florida. Another species, the Ghost Tiger Beetle, has virtually all-white elytra and lacks a distinctive central dark stripe down the length of the middle of the elytra. However, it is unrecorded from the southeastern United States.

Map 116 Moustached Tiger Beetle, *Ellipsoptera hirtilabris*.

Subspecies and morphological variants: No subspecies have been described.

Distribution and habitats: Confined to open patches of sand in pine woodlands, eroded sand hills, base of sand dunes, roadside ditches and dry, sparsely vegetated soil with white sand. From southeastern Georgia south over most of peninsular Florida except the marshy southern tip.

Behavior: Very fast runner but "freezes" in position when danger approaches. Flies up only at the last minute and then for a short distance where it seeks cover in sparse vegetation or the base of a tree.

Attracted to lights at night. Often found together with Scabrous Tiger Beetle and Highlands Tiger Beetle.

Seasonality: Adults active from May to November but mostly June and July, and the species has a 1- or 2-year life cycle.

Larval biology: Burrows are deep in stabilized sand soils usually away from vegetation. Larvae are active throughout the year.

Ecology and Behavior

Physical Habitats

There are at least seventeen distinctive habitats into which we can place almost all tiger beetle species the United States and Canada, such as sand dune, ocean beach, hardwood forest floor, and so on. Most tiger beetles are limited to a single habitat type, and only a few species, such as the Oblique-lined Tiger Beetle, occur in as many as six of the habitat categories. Larvae tend to be even more restricted to habitat type than do the adults. For both adults and larvae, however, each habitat type is made up of numerous components. These include physical, chemical and climatic qualities such as soil composition, moisture, temperature, and chemistry, vegetation cover, seasonality, as well as food supply. In addition, other habitat characteristics are important for providing mating and oviposition sites, and hiding places from parasites and predators. The ensemble of special adaptations that each species exhibits then allows or restricts it to a unique range and type of habitats.

Temperature is a critical and general part of the habitat for tiger beetles. Tiger beetles are *ectothermic*; that is, they are largely dependent on external sources of temperature to maintain their internal body temperatures so that they can be active. Because adult tiger beetles engage in much running and flying, they maintain internal body temperatures that are just below their lethal limits of 39°C. These high internal temperatures allow for maximum speed and movement. A sluggish adult is less likely to escape enemies, chase down mates, or capture prey. On the other hand, internal temperatures that are too high cause water balance problems, reduce gamete production, and affect general metabolism. We know, for instance, that different species of tiger beetles in the same locality can be active at and tolerate different maximum and minimum temperatures. These dissimilar capabilities in the same microhabitat may force species to be active at different times of the day and thus avoid competitors and predators. They may also explain some of the differences in geographical range of species.

Although structural features such as body size and body color can be important for adjusting internal temperatures, behavior is one of the most obvious adaptations used for thermoregulation. To regulate high internal temperatures during the day, adult tiger beetles initially extend their long legs (Plate 26.1) (stilting), placing their body above the thin layer of hotter air right next to the soil surface. As this layer of hot air broadens with the heat of the day, the beetles combine stilting with an inclined orientation of their bodies toward the sun, a position called sun facing that exposes only the front of the head to the sun's direct rays. When temperatures rise enough to make even these behaviors ineffective for controlling internal temperatures, the beetles avoid high temperatures by seeking out wet substrates, digging burrows in cooler substrates (Plate 26.2), or becoming inactive in shaded places. Running in and out of shaded areas (shuttling) (Plate 27.1) is another common temperature-controlling behavior. On cool mornings some species will crouch down to contact the warmer soil surface with the underside of their body (Plate 27.2, 28.1).

For larvae, placement of their burrows in the shade of vegetation or raised on chimney-like turrets (Fig. 7.1) also helps regulate internal temperatures. Extra deep burrows make it possible for some northern larvae to pass the winter below the frost line.

Most adult tiger beetles are active during the day, but some species of desert and subtropical habitats, on the other hand, are active at night, presumably at least partially to escape extreme daytime temperatures.

Figure 7.1 Larval Williston's Tiger Beetle (*Cicindela willistoni*) turret raised above the substrate surface. Photograph by D. L. Pearson.

However, species of genera such as the Night-stalking Tiger Beetle (*Omus*) are primarily nocturnal in temperate rain forests where daytime temperatures are relatively low. Other species forage actively during the daylight hours but mate or deposit eggs at night. Adults of some species that are normally active during the day are regularly attracted to lights at night.

Flooding is another important habitat factor, especially for the many species found in various water edge habitats. On the one hand, it attracts abundant prey for both adults and larvae and suitable egg-laying sites. However, frequent or extreme flooding may negatively impact tiger beetles, especially the relatively immobile larvae. Larvae of shoreline species such as the Eastern Beach Tiger Beetle and White-cloaked Tiger Beetle can survive when kept under water for as long as 6–12 days. Preliminary laboratory tests show that they are able to reduce their metabolism by as much as 90% by cutting back on cellular biochemical reactions that use oxygen. They may also be able to breathe air trapped in their closed tunnels. Even with these adaptations for surviving flooding, however, the larvae have lost valuable development time.

Dispersion is another adaptation for dealing with rapidly changing environmental conditions. For example, the *maricopa* subspecies of the Western Tiger Beetle is found in southwestern desert stream beds. Here it depends on moist sandy beds for keeping internal temperatures from going too high, finding mates and prey as well as for places to put its eggs into the soil. A single summer rainstorm can produce a flash flood that randomly scours sand in these streams down to bedrock in some areas and redeposits deep sand in other areas. A thriving population present for several years at one site can be gone in a day along with the entire local habitat. A favorable habitat appears just as suddenly somewhere else. Evidently by flying or running along the stream edge, surviving adults of this tiger beetle quickly locate new suitable habitat. The often random destruction of its habitat makes the advantages of this adaptation for dispersion obvious.

Seasonal Cycles

Most of the Common Tiger Beetles (tribe Cicindelini) display one of two distinct life cycle patterns. Adults of some species are active only in the summer. Others, meanwhile, have a split their activity period into two parts: the spring and again in the fall. The main difference between summer species and spring–fall species is in their overwintering stage. Whereas summer-active species die off as adults at the end of the summer and overwinter only in the larval stage, spring–fall active species overwinter as adults, but, if they require more than one year for development, they also overwinter in the larval stage the first winter. Although larvae can also show either pattern of seasonal activity, the more constant environmental conditions of their underground tunnels apparently enables them to be active throughout much more of the year than the more exposed adults.

There are some notable exceptions to these two patterns of annual adult activity, and they are mainly the result of peculiar weather factors in some areas. The Beautiful Tiger Beetle and Williston's Tiger Beetle have a spring–fall activity cycle in the northern part of their ranges but a summer activity in the southern part. The common forest species, Six-spotted Tiger Beetle that occurs throughout eastern North America, is active primarily in the spring and early summer everywhere in its extensive range. The Autumn Tiger Beetle of the southeastern United States is active only during the fall as are the Cazier's Tiger Beetle and subspecies of the Big Grassland Tiger Beetle and Black Sky Tiger Beetle in South Texas. The Ohlone Tiger Beetle is unique in North America by having a late winter activity season (February and March).

Although adults of most species of North American tiger beetles are active for only about 2–3 months during the spring, summer or fall, adults of some South Texas species (Nevada, Coastal, Gulfshore, and Cream-edged Tiger Beetles) are active from May into November, and adults of the Miami Beetle are active from May into October. This extended period of adult activity is most likely a result of two separate cohorts developing during the extended warm fall in these southerly locations.

In general, more spring–fall species are in the northern part of the continent, and they are the only type present at the northernmost locations. Summer-active species are the most common type in the south. But habitat type may also be involved. Summer species predominate in tidal areas, sandy beaches, and muddy inland habitats, whereas only spring–fall species occupy high-elevation alpine habitats.

Mating Behavior and Egg Laying

Adults of summer-active species begin reproduction soon after emergence from their pupal stages, but adults that emerge in the fall delay sexual activity until the following spring. A few observations suggest that males of some species emerge before the females. To begin the mating process, a male approaches a female in intermittent sprints similar to those used in pursuing prey. When he gets close enough, he leaps onto the back of the female, grips the sides of her thorax with his mandibles, and grasps her elytra with his middle and front legs (Plate 28.2). The male of most species, however, will often struggle with his target as she attempts to throw him off her back. Only if a male can success-fully hold on, does he have a chance of fertilizing the female's eggs.

Tiger beetles have several specialized structures to aid in successful mating. Males of almost all species of tiger beetles have the bottom surfaces of several tarsal segments of the front legs thickly covered with pads of hair-like setae (Fig. 2.5) that apparently aid in grasping and holding the female. His hind legs remain on the substrate, and he uses them to walk along as the female moves around. Females of many species have unique grooves and indentations (*coupling sulcus*) in the rear part of the sides of the thorax. These grooves provide purchase for the male's mandibles to hold on to the female. After mating, the male may hold on with his mandibles and legs and continue to ride the female for as long as several hours, a behav-ior called mate-guarding (Plate 29.1). This behavior serves to exclude other males from fertilizing her eggs.

Most of the time, tiger beetles will only engage in mating with other mem-bers of their own species. However, in some instances closely related species may overlap in geographic range and may mate with each other at times. These matings can result in hybrid offspring, as is observed between the Western Tiger Beetle and Twelve-spotted Tiger Beetle where they come in contact in parts of the Rocky Mountains. Occasionally males of one species will attempt to mate with a female of a distantly related species in a misguided effort that would be unlikely to result in genetically viable offspring (Plate 29.2).

After the egg is fertilized, the female injects it into the soil (*oviposi-tion*). To test a site before depositing an egg, the female uses her antennae and mandibles. If conditions are appropriate in terms of soil texture, salin-ity, slope, moisture, and temperature, she extends her ovipositor and then turns her body almost vertical (Plate 30.1). With abdominal thrusts and a digging-cutting action of the end of the ovipositor, she inserts the end of the ovipositor several millimeters to more than a centimeter below the surface and injects a single ovoid-shaped egg. She then removes the ovipositor and covers the hole so that no disturbance is obvious. Recent studies have found that females of some species oviposit while they are in adult burrows. She can oviposit ten to twenty eggs in a day in captivity, but few counts in the wild have been made. The emerging larvae construct their burrows at the site chosen by the female for oviposition.

The larvae go through three stages (instars) in which they molt into larger and larger bodies. The larval life ends when the third instar larva enlarges its burrow and turns into a resting stage called the pupa. In this stage, the adult structures grow and develop. Finally, the adult beetle emerges from the pupa to start the cycle over again. The length of time from egg to pupa is variable and depends on food availability, weather, and habitat characteristics. Most species have a 2-year life cycle, but some species develop in 1 year while others require 3 or more years.

Predation and Parasitism

For adult tiger beetles, the most important predators are robber flies (Plate 30.2), lizards and birds (Plate 31.1). Some birds (kestrels and flycatchers) and robber flies generally catch adult tiger beetles in the air during flight. Lizards and some bird predators like shrikes catch tiger beetles on the ground before they can fly away. In some areas spiders, scorpions, other arachnids, and predatory bugs can be important predators on adult tiger beetles. Adult tiger beetles commonly have attached to their legs and antennae the heads of ants (Fig. 7.2), which may indicate that groups of attacking ants regularly overpower and kill tiger beetles.

Enemies of larval tiger beetles include ground-foraging woodpeckers, ants, and provisioning wasps, but most important are attacks by parasitoid wasps and flies. While predators capture, kill and eat their prey, parasitoids instead lay their eggs on or near the prey, and the larval parasitoids then eat the prey item when they emerge from the eggs. Many species of the worldwide genera of ant-like wasps *Methoca* and *Pteromborus* (Tiphiidae) specialize in locating ground-dwelling tiger beetle larvae in their tunnels. These small, female wasps sting and paralyze the tiger beetle larva. Then the female deposits her egg on the immobilized larva, plugs the chamber, and fills in the top part of the tunnel with soil. The larval wasp hatches in 4–5 days, consumes the tiger beetle larva, and emerges as an adult wasp.

Another major parasitoid of larval tiger beetles are bee-flies of the genus *Anthrax* (Bombyliidae). Females of these flies hover over the tunnel entrance or settle on the ground nearby but just out of reach of the larva's mandibles.

Figure 7.2 Big Sand Tiger Beetle (*Cicindela formosa*) with ant head attached to antenna. Photograph by C. R. Brown.

The fly then turns her abdomen under and flips eggs at the tunnel entrance. The eggs that fall into the tunnel then roll to the bottom, where the fly larva soon hatches out. The parasitoid crawls onto the larval tiger beetle, attaches itself to the underside or top of the abdomen or thorax (Fig. 7.3), and stays there until the beetle reaches its pupal stage. During this stage, the tiger beetle is defenseless and the fly larva consumes it.

Other nonlethal parasites will attack adult tiger beetles, including mites in the family Trombidiidae. These small arachnids are typically bright red and immobile and can be seen attached to the legs or thorax of a tiger beetle. Although they are much more common in the warm tropics, these parasitic mites may infrequently occur on North American species as well (Plate 31.2).

Antipredator Defenses

Excellent vision to detect danger coupled with quick escape flights and fast running are the primary defenses for most adult tiger beetles. Chemical defenses such as benzaldehyde and cyanide are released against some predators, such as robber flies. Large body size deters smaller predators. Adults of most diurnal species spend the night protected in burrows or under detritus and vegetation. In some cases, adult tiger beetles will perch at night on low vegetation to escape ground predators notice (Plate 32). Most nocturnal species spend the day resting in these same types of refuges.

Other evidence suggests many tiger beetles produce sounds by rubbing body and wing parts together (*stridulation*), which may be used to distract or alarm predators. Sounds produced by pursuing predators, such as high-pitched ultrasounds used by bats in echolocating flying prey, may be perceived by the sensitive ears of tiger beetles to avoid them. These ears (*tympana*) are rare among beetles. On tiger beetles, they are located on the top of the abdomen and under the base of the elytra.

Figure 7.3 Parasitoid fly larvae attached to larval tiger beetle. Photograph by S. M. Spomer.

Camouflage is another widely observed antipredator mechanism. Adult color patterns of most tiger beetle species closely match the color of the soil substrate on which that species characteristically occurs. In addition, the three pairs of maculations on the elytra often serve to disrupt the shape and form of the tiger beetle. Other species have color patterns that mimic inanimate objects, such as pebbles or wood detritus, common in their habitat.

Bright orange abdomens that are revealed in flight when the elytra are opened keep several aerial predators away, but these warning or aposematic colors can involve much more of the body on some species. Most of these cases are linked with similarity in appearance to another noxious insect found in the same area. Some examples of possible mimicry include the Beautiful Tiger Beetle that looks like a brightly colored or obviously patterned female wasp in the family Mutillidae (called velvet ants). This velvet ant has a potent sting and runs on soil substrates in tiger beetle–like sprints. Some of these velvet ants even produce stridulation sounds and defense chemicals similar to their counterpart tiger beetles.

Larval tiger beetles rely primarily on a few behavioral responses to escape or avoid their enemies. Most commonly, they quickly retreat from the top of the tunnel to hide at the bottom. In addition, if a predator is able to grab a larva in its tunnel, the hooks on the back of back of the larva (Fig. 2.4) dig into the side of the tunnel and make it difficult for a predator

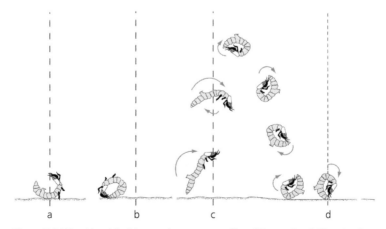

Figure 7.4 Wheel (tumbling) locomotion sequence of larval Eastern Beach Tiger Beetle (*Habroscelimorpha dorsalis media*) across smooth sandy beach. **a**. coiling and rolling backwards, **b**. straightening out and pushing off the sand into the air, **c**. rotating forward, **d**. landing on the sand and wheeling across the surface before the wind. Illustration by A. W. Harvey.

to physically pull the larva out. The depth of the tunnel, curves in its course, and hard clay, sandy, or rocky soil can hinder digging attempts by predators. As a last recourse, a larva will abandon a tunnel to escape an enemy, scuttle across the soil surface using its short legs and undulating body movements, and find a site in which it quickly excavates another tunnel.

Recently larvae of the Eastern Beach Tiger Beetle were found to escape predators by leaping somersaults and then rolling across hard-packed sand propelled by the wind. They can quickly move as far as 60 m using this wind-powered wheel locomotion (Fig. 7.4).

Larvae detect danger largely through vision. Studies have revealed a fine focusing visual acuity. The larval eyes are important for both locating potential prey items and accurately detecting danger. The larvae may also use vibrations through the ground to detect larger predators.

Competition

Competition occurs when one individual uses so much of a critical resource that another individual gets less than it could use. The critical or limiting resource can be space, egg-laying site, water, food, mates, or any number of other substances. Competing individuals can be of the same species engaging in intraspecific competition or of different species engaging in interspecific competition. Numerous studies around the world show that food is the most common resource for which tiger beetles compete, and thus feeding behavior becomes critical to understanding which tiger beetles can occur together.

Feeding Behavior

Tiger beetle larvae wait at the opening of their tunnels for arthropod prey to approach within striking distance. To capture the prey item, a larva rears backward up to half of its length out of the tunnel and grasps the prey with its mandibles. Larvae of some species dig cone-shaped depressions in this attack zone, apparently to increase the probability of capturing prey struggling to get out of the depression. Others construct turrets above the tunnel opening that function, at least partly, to provide an attractive landing spot for potential prey. When the prey item is successfully captured, the larva pulls it to the bottom of the tunnel and eats the digestible portions. Larvae attack and eat most types of prey except those with noxious chemicals and those too large to be quickly subdued and dragged to the bottom of the tunnel. After eating the digestible parts of the prey, the larva carries the indigestible remains to the mouth of the tunnel and throws them backward away from the tunnel opening.

Adults are visual predators and capture a wide variety of living arthropods. Though they have excellent sight, when they chase prey, they run so fast they can no longer see where they are going. Thus, adult tiger beetles use a pursuit pattern that involves active running interspersed with a pause-and-look behavior. Alternatively, the tiger beetle waits in a shaded area and ambushes approaching prey. The tiger beetle then grabs the prey item with its mandibles (Plate 33.1). Other adult tiger beetles frequently eat dead organisms (Plate 33.2) and occasionally fallen fruits.

Recently researchers discovered that, unlike most other insects that wave their antenna around as feelers to acquire information, tiger beetles hold their antennae rigidly and directly in front of them to mechanically sense their environments and avoid obstacles while running.

Adaptations Against Competition

If two species of tiger beetles compete for the same food, one early outcome is that the better competitor is so successful it captures most of the available prey. The second species of tiger beetle can no longer continue in the area and is forced to move to a new region or goes extinct. This reaction of the poorer competitor population to the dominant competitor is called competitive exclusion but is difficult to observe in the field. In most cases we see only the result and rarely witness the actual process of exclusion itself.

Another effect of competition that we can more easily observe is segregation of species in space and habitat. Victor Shelford, one of the founders of American ecology, reported that adult tiger beetle species on the southern shores of Lake Michigan occupied different microhabitats from water's edge to oak forest floor, so that two species rarely occurred together. Similarly, grassland tiger beetle species in the mountains of New Mexico each occur at different elevations. In the southwestern United States, tiger beetle species along sandy beaches of desert rivers used different parts of

Figure 7.5 Great Plains Giant Tiger Beetle (*Amblycheila cylindriformis*) using its mandibles to process a large prey item. Photograph by M. L. Brust.

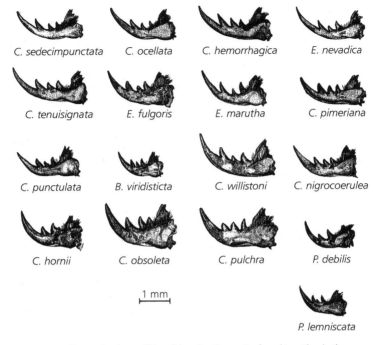

C. sedecimpunctata   *C. ocellata*   *C. hemorrhagica*   *E. nevadica*

C. tenuisignata   *E. fulgoris*   *E. marutha*   *C. pimeriana*

C. punctulata   *B. viridisticta*   *C. willistoni*   *C. nigrocoerulea*

C. hornii   *C. obsoleta*   *C. pulchra*   *P. debilis*

1 mm

P. lemniscata

Figure 7.6 Different-sized mandibles of tiger beetle species found together in the Sulphur Springs Valley of southeastern Arizona.

the beach. Species on saline flats in the Midwest use differential tolerance of pH and salinity to occur in these extreme habitats where many other species of tiger beetles cannot survive. In the few studies that include larval tiger beetles, spatial separation by microhabitat is even more pronounced than for the adults. This extreme separation of species in their larval stages may indicate that competition is potentially more important among larvae than among the adults.

However, there are many examples of tiger beetle species that use the same food and occupy the same microhabitat, but they usually are active at different times. In many regions, species with spring–fall active adults occupy the same microhabitat at a different season than species with summer-active adults. Different species in the larval stage in the same area also separate activity by season, thereby perhaps reducing competition. Daily activity periods may also be different. Nocturnally active species may thus potentially reduce competition from diurnally active species in the same habitat.

For species that occur together in the same microhabitat and at the same time, they may reduce competition by dividing up the prey. Everywhere it has been tested, mandible lengths of adult tiger beetles are highly correlated with prey size captured and eaten (Fig. 7.5). Small mandibles are not effective in grabbing or subduing large prey items. Large mandibles, in contrast, can handle large prey items, but they may be clumsier at processing small prey. Thus, tiger beetle species occurring together in the same microhabitat at the same time of year and hour of the day could still reduce the impact of competition for food if they each had different-sized mandibles (Fig. 7.6) and the associated differences in prey preference. Indeed, in most habitat types where food is not abundant, tiger beetle species occurring together tend to have different-sized mandibles.

Conservation

Many of the characteristics of tiger beetles that have generated interest among amateurs and professional biologists have also contributed to their increasing role in conservation studies. Most important among these characteristics are the ease with which many species can be found and identified in the field, their habitat specificity, and their value as indicators of undisturbed or natural habitats and of biodiversity. Also, because they have been well-collected and studied, their past and present distributions can be compared to evaluate historic trends of decline in range or abundance.

Tiger beetles, like other animals and plants that are endangered or threatened with extinction, can be protected under the U.S. Federal Endangered Species Act. This law prohibits human related adverse impacts to the species and their habitats, including collecting specimens. The Act also provides for recovery actions, such as funding to restore or acquire habitat. Prior to listing a species or population, the U.S. Fish and Wildlife Service must obtain information on the current distribution and abundance, potential or actual threats, and evidence of historic decline and current threats. The provisions of the Endangered Species Act also make it possible for private citizens or groups to submit a petition to list a species or population. The petition should provide new information on the status of the species or cite previous studies. The petition must be considered by the U.S. Fish and Wildlife Service and a response given within 90 days as to whether or not the species will be listed within two years. Such petitions have been the primary impetus for the listing of several species of tiger beetles.

We estimate that at least thirty-six (15%) of the named species and subspecies of tiger beetles in the United States and Canada are now so rare that they should be considered for inclusion on the U.S. Fish and Wildlife Service's List of Endangered and Threatened species. At present, only four of these are officially listed as threatened or endangered, and two others are being considered for listing.

Listed as endangered are the Ohlone Tiger Beetle (*Cicindela ohlone*) and the *lincolniana* subspecies of the Nevada Tiger Beetle (*Ellipsoptera nevadica*), otherwise called the Salt Creek Tiger Beetle. Listed as threatened are the Puritan Tiger Beetle (*Ellipsoptera puritana*) and the northeastern subspecies, *dorsalis*, of the Eastern Beach Tiger Beetle (*Habroscelimorpha dorsalis*).

Candidates for listing as threatened are the Coral Pink Sand Dune Tiger Beetle (*Cicindela albissima*) and the Highlands Tiger Beetle (*Cicindelidia highlandensis*).

The *abrupta* subspecies of the Hairy-necked Tiger Beetle (*Cicindela hirticollis*) and the *smythi* subspecies of the Lime-headed Tiger Beetle (*Habroscelimorpha chlorocephala*) are now considered extinct.

Conservation History and Concerns for Endangered and Threatened Tiger Beetles

Northeastern Beach Tiger Beetle
(*Habroscelimorpha dorsalis dorsalis*)

This subspecies was listed as threatened by the U.S. Fish and Wildlife Service in 1990 because of extirpation from nearly all of its range in the northeast and the lack of adequate protection of the Chesapeake Bay sites. In the early 1900s, this beetle occurred in large numbers along nearly all of the sandy beaches from central New Jersey to Massachusetts, but it declined rapidly throughout the early to mid-1900s coincidental with increased development and recreational activity along the Atlantic shoreline. By 1980 it was believed extinct in the northeast, but in 1990 a population was found on a well-protected beach on the island of Martha's Vineyard, Massachusetts. Population size at this site has ranged from several hundred to over a thousand. Subsequently two additional small populations were found, one at another beach on Martha's Vineyard and one, now possibly extinct, on the mainland of Massachusetts. Both of these satellite colonies may have been established from beetles which dispersed from the original Martha's Vineyard site.

Intensive surveys along both shorelines of the Chesapeake Bay of Virginia and Maryland in the 1990s confirmed the presence of this subspecies at over 100 sites, many with small numbers of adults. A few sites with large populations have been protected, but almost all of these sites are threatened to some degree by shoreline erosion and the construction of shoreline modifications (groins, bulkheads, rip-rap) that reduce the habitat quality for *Habroscelimorpha dorsalis dorsalis*. Indeed, surveys from 2009 to 2012 indicated this beetle already has disappeared from at least thirty of these sites. The recovery plan for this species calls for the permanent protection of additional sites within the Chesapeake Bay and the reestablishment of new populations by translocation in New England. One attempt at recovery of this subspecies was the reintroduction and establishment of a population at Sandy Hook, New Jersey. Here nearly 2000 larvae were translocated over a 4-year period. Unfortunately, the resultant adults did not survive, and the experiment was discontinued. However, another translocation using small numbers of larvae for several years from the Martha's Vineyard population to Monomoy Island, off the north tip of Cape Cod, resulted in the establishment of a new population of more than 1000 adults.

Puritan Tiger Beetle (*Ellipsoptera puritana*)

The historically disjunct range of this species included populations along the Connecticut River from New Hampshire to Connecticut and along the Chesapeake Bay shoreline in Maryland. In both areas, adults are found along narrow sandy beaches, and larvae occur with adults on the sandy beaches along the Connecticut River and on steep, high bluffs behind narrow bay beaches. The Puritan Tiger Beetle was listed as a threatened species in 1990 because of its disappearance from most of the New England sites and threats to the populations in Maryland. Only four populations remain in New England, three within a few miles of each other in Connecticut and one in Massachusetts. Evidently the species disappeared from other sites New England because of dams, channelization and other river edge impacts. Recent surveys in Maryland have identified two clusters of populations, one in Calvert County with populations at nine sites and the other with populations at eight sites around the mouth of the Sassafras River in Maryland's northern Eastern Shore. Total adult numbers in both population clusters have fluctuated twofold to threefold over the years but have generally declined from the early 1990s. The likely causes of this decline are shoreline erosion and human disturbances to the cliff and shoreline habitat.

Ohlone Tiger Beetle (*Cicindela ohlone*)

In response to a petition from biologists, this recently discovered species, was listed as endangered in 2001. At that time it occurred on fifteen small remnant coastal terrace grassland sites in Santa Cruz County, California. By 2012, it was present on only seven of these sites. Populations at these sites are small (less than 100 to several hundred individuals), and they are seriously impacted by encroachment of introduced grasses and weeds that have greatly reduced the open bare patches of soil required by this species. Cattle grazing and mountain bike activity at several sites are compacting soil and crushing some adults, but, on the other hand, they are also creating important open soil areas along the trails. Current studies are now gathering more details of the life history of the Ohlone Tiger Beetle. With this new information, better decisions can be made about management strategies needed to ensure its survival.

Coral Pink Sand Dune Tiger Beetle (*Cicindela albissima*)

The entire population of this species occurs within a 400-hectare portion of the 11-km long Coral Pink Sand Dunes in southern Utah. It was a candidate for listing as endangered in 1996, but instead it received protection from a Conservation Agreement initiated in 1998. Under this plan most of the adult and larval habitat is protected from potential impacts of off-road vehicles. Monitoring of this population over the past 15 years shows total population size ranges from 600 to 3000 individuals. The results of a recent Population Viability Analysis indicated this species is still at risk of extinction, and in 2012 it was proposed for listing as a threatened species by the U.S. Fish and Wildlife Service. In 2013, this proposal was withdrawn and a revised Conservation Agreement was implemented.

Highlands Tiger Beetle (*Cicindelidia highlandensis*)

This candidate species was described in 1985, and originally it was thought to occur at only a few sites on the Lake Wales Ridge in Highlands County, Florida. Subsequent surveys have identified over thirty sites where this species occurs, but they are all confined to central Florida and most are small in area and with small, marginal populations. Many of these areas are threatened by either the rapid conversion of land to housing developments, new citrus orchards or increased plant growth and succession resulting from fire suppression. Recent protection and management of several sites by the state of Florida and The Nature Conservancy have improved the status of this species. However, most sites remain vulnerable to threats from habitat fragmentation and genetic inbreeding.

Sacramento Valley Hairy-necked Tiger Beetle (*Cicindela hirticollis abrupta*)

This subspecies was known from only a few sites along short sections of the Sacramento and Feather Rivers in the northern Sacramento Valley of California. In 1998, an intense survey of all historic and other potential river edge sites on these two rivers found no adults or larvae. The primary cause of the apparent extinction of this tiger beetle population is likely due to the construction of the Oroville Dam on the Feather River in the 1960s. As is often the case with dam placement on rivers, the sand supply critical for maintaining flood plains upon which this species depends was cut off. Also, sustained high water levels have inundated the flood plain habitat for weeks at a time so that recruitment and development of larvae were curtailed.

Salt Creek Tiger Beetle (*Ellipsoptera nevadica lincolniana*)

The Salt Creek Tiger Beetle, a highly endemic subspecies of the Nevada Tiger Beetle, has long been known to have a limited range in the salt basin around Lincoln, Nebraska. Most of these sites have been lost to drainage and filling in recent years, and now because only three small populations remain, emergency listing as endangered was enacted in 2006. University of Nebraska biologists are monitoring populations and studying its biology in an effort to prevent its extinction.

Other Rare Tiger Beetles

Additional species, some even rarer than those officially listed as threatened or endangered, have not yet been listed either because they have not been considered by U.S. Fish and Wildlife Service and/or because reliable information on their current status is lacking. Here we present the most reliable information on their distribution and abundance as justification for their consideration for listing in the future. Over half of these species are in the states of California and Texas, and only four occur in eastern states.

Most of the rare California tiger beetles are southern coastal or inland species that have been severely impacted by habitat loss from the urbanization and development associated with population growth in the area. The *gravida* subspecies of the Hairy-necked Tiger Beetle (*Cicindela hirticollis*) historically ranged from Pt. Reyes near San Francisco south along the California coast into Mexico. Most records have been from the southern part of its range in the United States, from Los Angeles to San Diego. At present, it is believed to occur at fewer than ten widely separated sites from Pt. Reyes to San Diego. Most of these sites are on protected public lands, but the present patchy distribution and low abundance of this subspecies makes it vulnerable to extirpation or decline from even minor shoreline changes.

The Western Beach Tiger Beetle (*Cicindela latesignata latesignata*) formerly occurred in several counties along the southern California coast, but it is now gone from all but four of those sites. It does, however, still range south along the coast in Baja California where it is more common and widespread.

The Western Tidal Flat Tiger Beetle (*Eunota gabbii*) and the Western S-banded Tiger Beetle (*Cicindelidia trifasciata sigmoidea*) have distributions similar to that of the Western Beach Tiger Beetle in California and Mexico. These species are also limited to only a few sites in Southern California and are threatened by shoreline changes, habitat disturbances and the problems of small population size.

The Southern California form of the Senile Tiger Beetle, considered a subspecies, *Cicindela senilis frosti*, by some workers, has been found at only a few inland sites in recent years.

Cicindelidia hemorrhagica pacifica is considered a distinct subspecies by some workers because of its blue coloration and localized coastal beach habitat. Its former limited range has been greatly reduced and the few remaining populations are threatened by increasing recreational use of beaches.

Two California subspecies of the Oblique-lined Tiger Beetle (*Cicindela tranquebarica*) are in trouble. Most of the historic sites for the Southern California Oblique-lined Tiger Beetle (*C. t. viridissima*) have been lost to urbanization. It was found at only a few sites near Riverside 10–20 years ago, but it now may survive at only one site. The recently described Joaquin Valley Oblique-lined Tiger Beetle (*C. t. joaquinensis*) has disappeared from nearly all of its known historical sites in the San Joaquin Valley, and it now occurs at only four small patches of alkali sink habitat.

The Meadow Tiger Beetle (*Parvindela lunalonga*) had an historic range that included the San Joaquin valley north to the mountain meadows of the west slopes of the Sierra Nevada in northern California. Despite intensive searches over 40 years, it was only rediscovered in 2003 in a mountain meadow site in northern California. Most of the Central Valley sites have been eliminated by agricultural development and urbanization, but several populations have been found recently along irrigation ditches west of Stockton.

Three other rare tiger beetles are found in the Pacific Northwest. The recently described Mt. Ashland Night-stalking Tiger Beetle (*Omus cazieri*) is known only from the vicinity of Mt. Ashland in Oregon. The Bruneau Sand Dune Tiger Beetle (*Cicindela waynei*) is endemic to a small dune field in southwestern Idaho, and its population numbers are small. This species is apparently being impacted by plant succession on the dunes and possibly by overzealous collecting. The Columbia River Tiger Beetle (*Cicindela columbica*) was considered for listing by the U.S. Fish and Wildlife Service because it no longer occurs along most of its former range on Columbia River (Washington and Oregon) and Snake River (northwestern Idaho) beaches. Officials, however, in 1990 decided that it did not warrant listing because some of the remaining populations were in protected areas along the Salmon River in Idaho.

Texas is second to California in the number of declining tiger beetles. The *smythi* subspecies of the Lime-headed Tiger Beetle (*Opilidia chlorocephala*) probably is extinct. There is only one known collection site from the south

Texas coast in 1912. The South Texas Giant Tiger Beetle (*Amblycheila hoversoni*), the *subtropica* subspecies of the Black Sky Tiger Beetle (*Cicindelidia nigrocoerulea*), the *neojuvenilis* subspecies of the Big Grassland Tiger Beetle (*Cicindelidia obsoleta*), and the Velvet Tiger Beetle (*Dromochorus velutinigrens*) all have limited ranges in south Texas and are rarely recorded. Most of this area is experiencing rapid growth and agricultural development that threaten these naturally rare tiger beetles. Cazier's Tiger Beetle (*Cicindelidia cazieri*) has been found at only a few localities in the Rio Grande Valley of south Texas in its preferred limestone outcrop habitats. It may be more widespread, but it is so difficult to detect and its adult activity period so brief, the actual status level remains uncertain. Recent studies of populations of the *olmosa* subspecies of the Nevada Tiger Beetle (*Ellipsoptera nevadica*) along the south Texas coast indicate they may represent a subspecies distinct from the New Mexico populations. Only three sites with small populations are known in Texas, suggesting this form may be at risk. In New Mexico there are more populations spread over a broader area, and some of them are in protected areas.

The *funaroi* subspecies of the Williston's Tiger Beetle (*Cicindela willistoni*) and the *albilata* subspecies of the Glittering Tiger Beetle (*Cicindela fulgoris*) are restricted to localized areas in New Mexico. Both subspecies are under pressure by water level disruptions that are altering the habitat.

Both subspecies of the Riparian Tiger Beetle (*Eunota praetextata*) have disappeared from most of their historic sites along southwestern desert rivers in the last 30 years. Their disappearance can be traced to dams and irrigation that have dried or severely disrupted river flow and caused loss of essential desert riparian habitat. However, following recent restoration of the Salt River in the Phoenix area, populations of the Riparian Tiger Beetle have reappeared. Their rapid recolonization indicates they probably dispersed from small moist refugia on lower parts of these rivers.

Two endemic tiger beetles from the area of the Willcox Playa in southeastern Arizona, the *citata* subspecies of the Nevada Tiger Beetle (*Ellipsoptera nevadica*) and the *sulfontis* subspecies of the Williston's Tiger Beetle (*Cicindela willistoni*), have always had a limited distribution and abundance. These subspecies now may be threatened by the depletion of ground water from their limited habitat as a result of the dramatic increase in irrigation for agriculture in the area. The *yampae* subspecies of the Festive Tiger Beetle (*Cicindela scutellaris*) is endemic to a small dune area of Moffat County in Northwestern Colorado. Records over the years suggest that the remaining populations are small, perhaps due to plant overgrowth of the dunes and agricultural activity. The *corpuscula* subspecies of the Hairy-necked Tiger Beetle (*Cicindela hirticollis*) has disappeared from almost all of its former range on the Gila and Colorado Rivers of Arizona. It may now exist only on the Virgin and Green Rivers of Utah.

In the eastern United States, four other species are considered so localized and declining that they may be in danger. Among them, the Miami Tiger Beetle (*Cicindelidia floridana*) was first discovered in 1934 and described as a variety of the Eastern Pinebarrens Tiger beetle (*Cicindela abdominalis*). It then disappeared and was not rediscovered until 2007. With more study of its morphology and biology, it was recently elevated to its own species. It now remains in tiny populations on three small contiguous pine rockland habitats in south Miami. Its total distribution occupies less than 4 hectares. Because of this extremely restricted distribution and serious impact by encroaching exotic vegetation, the Miami Tiger Beetle is arguably the rarest and most endangered of U.S. tiger beetles.

The *hentzii* subspecies of the Eastern Red-bellied Tiger Beetle (*Cicindelidia rufiventris*) is restricted to shrubby hills south of Boston and is found at only a few sites in small numbers. The Olive Tiger Beetle (*Microthylax olivacea*) is a Cuban species that apparently colonized the outer Florida Keys in the middle of the last century. It occurred only within a localized area on the Keys and has not been found there in several decades.

The *consentanea* subspecies of the Northern Barrens Tiger Beetle (*Cicindela patruela*) has declined dramatically from its former range throughout the Pine Barrens of New Jersey and Long Island, New York, to southeastern Pennsylvania, Delaware, and Maryland. In recent years it has disappeared from all of these areas except for a few highly localized populations in the New Jersey Pine Barrens. The decline of this subspecies is attributed to fire suppression and habitat alteration. Although historically much more widespread, the nominate subspecies of the Northern Pine Barrens Tiger Beetle also appears to have disappeared from much of its former range.

Several other tiger beetles were previously considered for listing. Among these, the Cobblestone Tiger Beetle (*Cicindela marginipennis*), the St. Anthony Dune Tiger Beetle (*Cicindela arenicola*), and the *maricopa* subspecies of the Western Tiger Beetle (*Cicindela oregona*) were found to be more common and widespread than previously thought and thus do not merit listing at this time. The three subspecies (*barbaraannae, petrophila,* and *viridimonticola*) of the Limestone Tiger Beetle (*Cicindelidia politula*) from west Texas and southern New Mexico were candidates for listing by U.S. Fish and Wildlife Service because of their limited distributions and threats of over collecting. Specimens were reportedly sold for $1500 each. Recent fieldwork, however, shows that *barbaraannae* is much more widely distributed than thought and that *viridimonticola* may not represent a distinct population. The last subspecies, *petrophila,* may be the rarest of these and is largely restricted to the Guadalupe Mountains.

Summary

It is evident that the publication of the first edition of this field guide generated increased interest in tiger beetles. An unforeseen consequence seems to be increased collecting, trading and selling of tiger beetles, especially the rarer ones. While there is little evidence that over collecting has to date seriously impacted any tiger beetle species, increased collecting of rare and localized species may put these forms at greater risk, decline or extinction. For that reason, we urge tiger beetle enthusiasts to limit numbers of specimens they collect. Workers can also help in the conservation of these and many other species by noting new information on distribution, population declines, and ecological and habitat limitations of adults and larvae. This information should be published or made available to appropriate workers, land managers of government agencies.

Observing and Studying Tiger Beetles

Field Observations

In addition to their beautiful colors, intricate designs, and usefulness as surrogates for understanding how to protect habitats, tiger beetles also have become popular because they can be so easily found and observed in the wild. Although wary and easily flushed by normal movements, they can be approached to within a meter or less by slow, careful movements. This will permit close observation of various behaviors or close-up photography. In the field, adult tiger beetles will become accustomed to you if you remain motionless or make a very slow and smooth approach. They will return to normal behavior within a few minutes after having been disturbed, and then you can watch them at close quarters. However, some species are more difficult to approach because of the thick mud, wet rocks, or other slippery substrate surfaces of their habitats. Close-focusing binoculars (generally less than 6 power) help make possible detailed observation of these more elusive species. Species active at dusk or during the night sometimes permit themselves to be observed with a flashlight if you do not shine the light beam directly on the beetles. However, some species are disturbed by any moving light source, and their behavior may be abnormal in these circumstances. The best alternative for accurate observations of nocturnal tiger beetles may be by using a night or starlight vision scope, binoculars or monocular. These optics have a light-sensitive illumination screen. The screen enhances extremely small sources of light that converts an otherwise dark world into ghostly illuminated objects. In total darkness, however, they will not function well.

For population and behavioral studies, beetles can be caught with a net, marked with spots of nontoxic colored paint (magic markers or more preferably paint pens with a xylene solvent, such as DecoColor ©) or numbered "bee" tags glued to the thorax or elytra, and then released. Unfortunately, because of their burrowing habits and extreme environments, the beetles usually end up scraping off the tags and paint within a few days. For longer-term studies of several weeks, it may be necessary to clip off a small tip of elytra or scratch off distinctive patches of white hairs on the undersides of hairy species. Estimates of population size are most easily made by directly counting numbers of beetles in particular areas in addition to various mark–recapture methods.

Field observations can also produce valuable information on the types and extent of predation on adults. Alternatively, you can use artificial models. Sturdy paper models of tiger beetles or dried tiger beetles themselves can be tied to string and attached to the end of a pole. With a little practice, these baits can be presented to predators in the field such as robber flies and lizards. Data on predator preference then can be gathered quickly. Some predatory birds, such as shrikes, can be entrained to come to insects presented to them under their perches. Choice preferences then can be monitored.

Larval tiger beetles quickly accommodate to a nearby observer, and some will accept prey offered to them on the end of fine forceps, but, as with adults, your movements must be very slow. Because of their burrow-dwelling habits, tiger beetle larvae are excellent subjects for studies of development, life cycles, parasitism, mortality, and, eventually, adult emergence.

It is difficult, however, to study natural feeding behavior or antipredator responses of larvae because of the infrequency of these events. Survival rates and other more long-term studies can be aided by marking active larval tunnels with numbered golf tees, roofing nails, or aluminum nursery tags. By coincidentally recording climatic conditions or food availability, the effect of these factors on larval population dynamics can be correlated. Census data through subsequent seasons or even years can then yield valuable results.

Laboratory Studies

Both adults and larvae can survive in a covered terrarium with adequate light sources, soil moisture and the right substrate. Adults are more difficult to maintain because of their attempts to escape from the terrarium and because it is hard to keep a proper balance of temperature and moisture for them. It is especially important to provide adults with a source of free moisture so they can drink frequently, but chambers must also have good air flow to prevent the buildup of fungus. Adults will forage on their own in a terrarium, but the prey should not be able to fly so that they do not all accumulate on the bottom of the cover overhead and out of reach of the adult tiger beetles. Ideal prey for both adults and larvae are adults and larvae of flour beetles, small crickets, vestigial winged fruit flies, or an assortment of field caught insects. Mating behavior and oviposition are more easily observed in a terrarium than in the wild. Larvae are much easier to raise and manipulate in terraria than are adults, and larvae can also be studied by placing them in individual plastic rearing tubes with soil. Studies of the effects of factors like moisture, temperature and food levels on larval development, survival, and adult fertility are most feasible in the laboratory.

Collecting Adults

If specimens are needed, many techniques have been developed to collect these generally wary beetles. Be aware, though, that permits for collecting insects, even if they are not listed as rare or endangered species, on state and federal lands are becoming more widespread. To obtain these permits, usually a written proposal and scientific justification are required along with the application. Even on private property, however, at least oral permission to enter and collect should be requested from the land owner.

Some collectors use a killing jar with a chemical such as ethyl acetate or similar solvent. But these chemicals must be recharged frequently, they are often highly flammable, and accumulation through the skin or breathing can impact the collector over a long period. Because tiger beetle specimens are so resistant to color changes, a safer and equally effective method is to collect directly into vials of 70–80% alcohol (ethyl alcohol or rubbing alcohol). The specimens can then be removed later for study and drying, or they can be stored long term in the vials with data labels. Glass vials with neoprene stoppers are usually more leakage resistant than vials or containers with screw-on lids. Nasco Whirl-Pac© plastic bags are especially easy to use for storing specimens in the field.

Using a standard insect net is the most common technique for capturing tiger beetles. Smooth slow movements by the collector are a must because tiger beetles quickly react to sudden, fast movements. Not only can the movements of the collector's body frighten the tiger beetles away, but even shadows moving across the substrate toward the beetle can elicit escape behavior. Once within net range, the beetles can be captured by slapping the net quickly onto the ground surface or by flushing the beetle up and then quickly sweeping the net just above the ground surface.

Pit fall traps provide an alternative collecting technique, particularly in more open habitats and for collecting flightless or nocturnal species. Cans or plastic cups can be buried in the ground with the lip of the opening flush with the substrate surface. Tiger beetles running in this habitat may fall into the cup. A funnel fitted into the top of the can or cup will make it less likely for individuals to escape back out the opening or for insectivorous predators to raid the trap before you return. Soapy water made with non-perfumed soap (smells can affect insect movements) will efficiently trap the insects by reducing the surface tension of the water. These traps must be checked at frequent intervals (at least once a day), however, as the insects will quickly die and rot. Because small vertebrates and many types of invertebrates can be trapped in these pit falls, legal permits may be required to use them. Placing the pitfalls at the edge of water or in natural runways such as at the base of boulders will enhance the probability of capture. In open areas, tall (15-cm) plastic lawn edging can be placed vertically in the ground in long strips between pitfalls to serve as barriers and direct beetles to run along its edge. An array of pitfalls with plastic walls extended in the shape of an "X" will capture beetles running in almost any direction. A few collectors have successfully baited the traps with decaying meat to attract some species of tiger beetles.

Night lights and flashlights can be employed to capture some species of tiger beetles. Portable black lights that shine ultraviolet light or full spectrum lights can be placed on or in front of a large white sheet. Placed vertically, the sheet reflects the light farther, and more insects are likely to be attracted. If the sheet is placed flat on the substrate, it is often easier to see more of the approaching tiger beetles, some of which may not come close to the light. Mantle lanterns, flashlights, or even the incandescent lights of an all-night store or gasoline station can attract tiger beetles. The beetles are usually very active at the light, flying, running and often pursuing prey species also attracted to the lights. You will have to use your fingers to catch them. Occasionally, diurnal species disturbed from their resting sites at night will fly out and come to lights, to which they are otherwise not normally attracted. Walking back and forth at night through grass or other vegetation may scare numerous specimens to a nearby light.

Sticky traps with petroleum-based chemicals are available under trademark names like Tanglefoot© and Tacktrap©. A thin layer of this extremely sticky substance can be painted on vegetation or onto plastic strips that are then placed on or tacked to the appropriate substrate. Alternatively, these substances are available already applied to standard squares of cardboard or thin plywood. When the insects run onto this substance, they are instantly immobilized. This method is most useful for collecting samples of tiger beetles and their prey over a period of several hours to a day. The main drawback of the technique is that the specimens are extremely difficult to clean. Water and alcohol will not work. Only solvents such as gasoline or paint thinner will do the job. A patch of clear plastic can be cut to cover the area of the sticky trap and placed on to the adhesive surface as a cover. The traps can then be handled and stored without the adhesive being spread to fingers, doorknobs, and steering wheels. Also, the plastic layers can be placed under a dissecting microscope and the trapped specimens counted, measured, and identified through the protective clear plastic covering.

Collecting Larvae

There are several ways to collect larval tiger beetles from their tunnels. First, you must master a search image for finding the almost perfectly round holes that these creatures build (Fig. 2.2). Often holes occur together in large numbers but in a confined area. Most are on flat and somewhat moist soil surfaces, but for some species they are on vertical banks. Some species plug the burrow entrances with soil in response to warm dry periods, and then they are impossible to see.

The most reliable technique for obtaining larvae is digging them from their burrows. First, place a thin blade of grass or a flexible straw down the tunnel, being careful to follow its curves and bends. Then using a garden towel or large spoon, dig a pit beside the tunnel, carefully pulling loose dirt out of the bottom of the excavation as you proceed. Cautiously remove the dirt along the larval tunnel using the grass or straw as a guide, and let the dirt drop into the bottom of the excavated pit. When you come to the end of the tunnel where the straw stops, very carefully remove the dirt to expose the larva. It will usually drop into the bottom of your pit where it is easily caught if you have kept the falling dirt from accumulating. Care must be taken to avoid injuring the larva with the digging instrument. This method is complicated by rocky ground and larval burrows that have sharp bends or turns in them.

"Fishing" is another alternative. With a little practice, this method is faster, more efficient, and less destructive to the habitat than the digging method. However, it does not work for all species, especially those with deep or sharply curved burrows. Use a blade of grass or a thin stick or straw with a bent "hook" at the end. Lower it into the larval tunnel until you meet resistance and then quickly but smoothly pull the straw up out of the tunnel's opening. Frequently, the larva's reaction to an intruding item is to bite it. If it holds on with its mandibles, it can be pulled out of the tunnel and caught.

Blocking the larval tunnel is a third option. Locate an active larval tunnel, and then place the blade of a knife or spoon into the substrate at an angle just below the tunnel opening but without penetrating the tunnel. Wait for the larva to move up to the tunnel mouth (probably several minutes) to begin foraging and then quickly push the knife or spoon into the tunnel below the larva so it cannot retreat. The larva can be quickly extracted from the upper part of the tunnel. The obvious drawbacks to this method include risk of cutting the retreating larva in half if your timing is bad, and the considerable patience needed to wait for the larva to return to the tunnel opening.

Preparing and Storing Specimens

Most of the techniques for preservation and storage of tiger beetles are similar to those for other insects, and many details are found in entomological techniques books. Three methods for long-term storage of tiger beetle specimens are widely used. For adult specimens, the most common method is piercing the front part of the right elytron with an insect pin (size No. 1 or 2 are best for this method) and sliding the specimen up toward the top of the pin. However, the beetle should be placed far enough down from the top of the pin so that there is room to hold the pin without touching the specimen. Use of a pinning block will ensure proper placement of the specimen and labels on the pin. The advantage of this method is that after the specimen has dried, all aspects of the beetle can be studied in detail without handling the specimen itself. A label or several labels can then be placed on the same pin under the specimen. These labels should be as detailed as possible indicating country, province or state, distance to nearest permanent settlement, latitude/longitude, elevation, date, habitat of the collected specimen, and collector's name. Labels should never be removed from the pin. The disadvantages of this system are few but include the hole produced by the pin through the body and the resultant destruction of some of the cuticle as well as internal organs. Coated steel insect pins with nylon heads are perhaps the optimum type to use.

Alternatively, glue can be placed on the underside of a specimen that is then mounted on a thin rectangle or triangle glue board cut or punched from an index card. An insect pin is put through the exposed end of the cardboard, and labels can be placed on the pin. In this manner, no damage is done to the specimen by the pin. This method also enhances the facility with which specimens can be photographed. The disadvantages include the inability to easily examine the underside of the specimen. A solvent must be applied to dissolve the glue, a procedure that involves additional handling of the fragile specimen. In addition, specimens are prone to fall off of the cardboard if inadequate glue is used.

Adult tiger beetle specimens often have considerable fat internally that can leach to the exterior over time. This fat deposit can discolor the body surface of dried specimens and clump the hair-like setae together. Many collectors place specimens in a solvent, such as Hexane or acetone, for several days to remove much of the lipid content. This process can be done before or after pinning the specimens. As with all solvents, great care should be taken to have adequate ventilation and avoid contact with your skin. They are also generally highly flammable. The color and setae of most specimens cleaned in this manner are not affected. However, cleaning specimens also can remove useful information about the habitat and natural history of a specimen. For instance, sand grains imbedded in the thick setae can provide habitat clues for subsequent researchers. Also, application of chemicals can affect subsequent analysis of molecular structure. If the cleaning is done on pinned specimens, make sure that the ink used for the label information is not soluble in these solvents.

No matter which technique you use for storing dried specimens, they must be constantly protected from warm, moist conditions to avoid fungal growth. They must also be kept in boxes, drawers, or containers that can be tightly sealed to prevent the entry of dermestid beetles and other arthropods that feast on cuticle and dried organs. Some collectors keep moth balls or other anti-insect chemicals in the storage containers, but these chemicals may have serious health effects on humans who breathe their fumes. Alternatively, the containers can be regularly placed in a deep freezer for several days. Be aware, however, that some of the arthropod pests that feed on dead insect bodies can accommodate to the freezing if the temperatures do not go low enough or if they are lowered too slowly.

Another useful technique is to store adult specimens in vials of 70–80% alcohol (ETOH) or rubbing alcohol (Isopropyl Alcohol). Vials with stopper tops are generally preferable to leak-prone screw top vials. A data label printed on high-quality (acid-free) paper with ink that will not run or dissolve in alcohol over long periods should be placed inside the vial. We have found that the full information of place, date, habitat, collector, and so forth should be on this label. Coded or abbreviated data often become separated from the notebooks or computer programs with the actual data. An additional label or code can be attached to the outside of the vial if desired. These labels on the outside of the vial often become removed or misplaced, so a label on the inside of the vial is mandatory. An advantage of this method is that the specimens' external characters and internal organs are preserved for later dissection and study. If too many specimens are placed in a vial, the crowding will result in dilution of the alcohol and inadequate preservation, separation of the tips of the elytra after pinning, and occasionally staining of colored surfaces.

Larvae should be kept alive in vials of substrate or other baffles until they can be preserved properly. They must not be placed with other larvae, or they can cannibalize each other. Larvae should then be placed directly into a solution of 10% formalin that has just been brought to a boil and removed from the heat. NOTE: Use a ventilation hood as formalin is toxic. Keep the larvae in the solution for about 30 minutes. Remove the larvae from the solution and soak them in water for 5 hours to rinse off the remaining formalin. The larvae together with their data labels can then be stored permanently in vials of 70% alcohol (ETOH). An alternative method is to drop larvae into boiling water, then transfer to 95% alcohol for long-term storage.

Recognizing the increasing use of specimens for molecular analysis and DNA studies, many museums and institutions are now developing frozen tissue collections. Alternatively, long-term storage (10–20 years) in alcohol at 20°C maintains DNA quality adequately for many analyses, provided that the alcohol is changed periodically when it discolors due to the dilution of fats and other chemicals in the bodies of the beetles. However, whenever possible it is preferable to keep ethanol preserved beetles in the refrigerator or freezer as this slows down the degradation of DNA.

For some DNA studies, the tiger beetles must be collected alive, frozen, or placed into either absolute or 96% ethanol or desiccation vials with silica gel. It is important that dehydration of the specimen is fast, and if silica gel is used, it must be in its blue (high hygroscopic) color state. Only a single specimen should be collected into each vial as they may break after desiccation and their body parts become so intermixed that DNA results cannot properly be attributed to a single specimen or species. Many chemicals can degrade DNA, so specimens captured in killing jars and those cleaned in solvents often are not useful for these molecular studies.

Photography

Just as for birds, butterflies, and dragonflies, the collection of tiger beetle specimens is becoming less common. Most collecting is now reserved for specific types of studies in which tissues, cells, and structures can only be observed and studied using specimens. For studies of behavior, ecology, and range extensions (and contractions), or for the pure pleasure of recording the beauty and intricacies of tiger beetle form, color, and pattern, photographs are more and more the preferred method.

Becoming a successful tiger beetle photographer, however, means developing skills not only in the techniques of camera use, but also in nature study. Patience, and more patience, is often the difference between an adequate shot and a drop-dead photograph. Learn all you can about your subject and its environment before you try taking pictures. How flighty is the species? If it does fly, will its flight path take it to a predictable part of the environment? How can you best stalk your subject, bearing in mind that this will include arriving in the spot that will allow you to take the best advantage of the shallow depth of field that close-up photography affords? In extreme cases, nonpurists will even capture the insect, cool it down in a refrigerator or with ice from a field pack for 15 minutes, and then release it into a controlled photographic situation, where the beetle's slowed activity allows more critical focusing and sharper images.

Type of camera and lens, lens aperture, ISO rating, shutter speed, the use of one or more flashes, recording format (typically RAW vs JPEG), and many other alternatives comprise only part of the array of decisions you will need to make. Whatever your decisions, there will be one constant element in tiger beetle photography: it will eventually involve a lot of crawling and lying on the ground. You will soon be selecting equipment that adapts well to this kind of behavior.

Whatever medium you choose to capture your tiger beetle images, your knowledge and appreciation of these marvelous creatures will be immensely enhanced as a result.

Selected Bibliography

Although for information on tiger beetles we relied on many articles, especially from the journals *Cicindela* and *Coleopterists Bulletin*, we list below the more general ones that we most often used for taxonomic, distributional, biological and conservation information of tiger beetles from the United States and Canada.

Acorn, J. 2001. *Tiger Beetles of Alberta: Killers on the Clay, Stalkers on the Sand.* University of Alberta Press, Edmonton, Canada, 120 pp.

Bousquet, Y. 2013. Catalogue of Geadephaga (Coleoptera, Adephaga) of America, north of Mexico. ZooKeys 245: 1–1722.

Boyd, H. P. and Associates. 1982. *Checklist of Cicindelidae: The Tiger Beetles.* Plexus Publ. Marlton, New Jersey, 31 pp.

Cassola, F. and D. L. Pearson. 2000. Global patterns of tiger beetle species richness (Coleoptera: Cicindelidae): their use in conservation planning. Biological Conservation 95: 197–208.

Cazier, M. A. 1954. A review of the Mexican tiger beetles of the genus *Cicindela* (Coleoptera, Cicindelidae). Bulletin of the American Museum of Natural History 103: 231–309.

Choate, P. M., Jr. 2003. *A Field Guide and Identification Manual for Florida and Eastern U.S.: Tiger Beetles.* University Press of Florida, Gainesville, Florida, 197 pp.

Duran, D. P.and R.A. Gwiazdowski. 2015. Systematic revision of Nearctic Cicindelini (Coleoptera: Carabidae: Cicindelinae): Re-evaluating Rivalier's taxonomy.

Erwin, T. L. and D. L. Pearson. 2008. *A treatise on the Western Hemisphere Caraboidea (Coleoptera): Their classification, distributions, and ways of life. Vol. II (Carabidae— Nebriiformes 2—Cicindelitae).* Pensoft Publishers, Sofia, Bulgaria, 365 pp.

Freitag, R. 1965. A revision of the North American species of the *Cicindela maritima* group with a study of hybridization between *Cicindela duodecimguttata* and *oregona.* Quaestiones Entomolgicae 1: 87–170.

Freitag, R. 1999. *Catalogue of the Tiger Beetles of Canada and the United States.* NRC Research Press, Ottawa, Canada, 195 pp.

Gwiazdowski, R. A., S. Gillespie, R. Weddle and J. S. Elkinton. 2011. Laboratory rearing of common and endangered species of North American tiger beetles (Coleoptera: Carabidae: Cicindelinae). Annals of the Entomological Society of America 104: 534–542.

Knisley, C. B. 2011. Anthropogenic disturbances and rare tiger beetle habitats: benefits, risks, and implications for conservation. Terrestrial Arthropod Reviews 4: 41–61.

Knisley, C. B. and T. D. Schultz. 1997. *The Biology of Tiger Beetles and a Guide to the Species of the South Atlantic States*. Virginia Museum of Natural History, Special Publications No. 5., Martinsville, Virginia, 209 pp.

Larochelle, A. and M-C. Larivière. 2001. Natural history of the tiger beetles of North America north of Mexico. Cicindela 33: 41–122.

Leonard, J. C. and R. T. Bell. 1999. *Northeastern Tiger Beetles: A Field Guide to Tiger Beetles of New England and Eastern Canada*. CRC Press, Boca Raton, Florida, 176 pp.

Naviaux, R. 2007. *Tetracha* (Coleoptera, Cicindelidae, Megacephalina): Révison du genre et descriptions de nouveaux taxons. Mémoires de la Société entomologique de France 7:1–197.

Pearson, D. L. 1988. Biology of tiger beetles. Annual Review of Entomology 33:123–147.

Pearson, D. L., T. G. Barraclough and A. P. Vogler. 1997. Distributional maps for North American species of tiger beetles (Coleoptera: Cicindelidae). Cicindela 29: 33–84.

Pearson, D. L. and F. Cassola. 1992. World-wide species richness patterns of tiger beetles (Coleoptera: Cicindelidae): Indicator taxon for biodiversity and conservation studies. Conservation Biology 6: 376–391.

Pearson, D. L. and A. P. Vogler. 2001. *Tiger Beetles: The Evolution, Ecology, and Diversity of the Cicindelids*. Cornell University Press, Ithaca, New York, 333 pp.

Putchkov, A. V. and E. Arndt. 1994. Preliminary list and key of known tiger beetle larvae (Coleoptera, Cicindelidae) of the world. Mitteilungen der Schweizerischen Entomologischen Gesellschaft 67: 411–420.

Rivalier, E. 1954. Démembrement du genre *Cicindela* Linné, II. Faune américaine. Revue française d'Entomologie 17: 217–244.

Wallis, J. B. 1961. *The Cicindelidae of Canada*. University of Toronto Press, Toronto, Canada, 74 pp.

Wiesner, J. 1992. *Checklist of the Tiger Beetles of the World*. Verlag Erna Bauer, Keltern, Germany, 364 pp.

Willis, H. L. 1967. Bionomics and zoogeography of tiger beetles of saline habitats in the central United States (Coleoptera: Cicindelidae). University of Kansas Science Bulletin 47: 145–313.

Checklist of the Tiger Beetles
of the United States and Canada

Genus *Amblycheila,* Giant Tiger Beetles

- ____ 1. *Amblycheila baroni,* Montane Giant Tiger Beetle
- ____ 2. *Amblycheila schwarzi,* Mojave Giant Tiger Beetle
- ____ 3. *Amblycheila cylindriformis,* Great Plains Giant Tiger Beetle
- ____ 4. *Amblycheila hoversoni,* South Texas Giant Tiger Beetle
- ____ 5. *Amblycheila picolominii,* Plateau Giant Tiger Beetle

Genus *Omus*, Night-stalking Tiger Beetles

- ____ 6. *Omus dejeani,* Greater Night-stalking Tiger Beetle
- ____ 7. *Omus submetallicus,* Lustrous Night-stalking Tiger Beetle
- ____ 8. *Omus californicus,* California Night-stalking Tiger Beetle
  - ____ *O. c. californicus*
  - ____ *O. c. angustocylindricus*
  - ____ *O. c. intermedius*
- ____ 9. *Omus audouini,* Audouin's Night-stalking Tiger Beetle
- ____ 10. *Omus cazieri,* Mount Ashland Night-stalking Tiger Beetle

Genus *Tetracha*, Metallic Tiger Beetles

- ____ 11. *Tetracha carolina,* Carolina Metallic Tiger Beetle
- ____ 12. *Tetracha floridana,* Florida Metallic Tiger Beetle
- ____ 13. *Tetracha impressa,* Upland Metallic Tiger Beetle
- ____ 14. *Tetracha virginica,* Virginia Metallic Tiger Beetle

Genus *Cicindela*, Temperate Tiger Beetles

- ____ 15. *Cicindela longilabris,* Long-lipped Tiger Beetle
- ____ 16. *Cicindela repanda,* Bronzed Tiger Beetle
  - ____ *C. r. repanda*
  - ____ *C. r. novascotiae*
  - ____ *C. r. tanneri*
- ____ 17. *Cicindela duodecimguttata,* Twelve-spotted Tiger Beetle
- ____ 18. *Cicindela oregona,* Western Tiger Beetle

_____ *C. o. oregona*
_____ *C. o. guttifera*
_____ *C. o. maricopa*
_____ *C. o. navajoensis*
_____ 19. *Cicindela depressula*, Dispirited Tiger Beetle
_____ *C. d. depressula*
_____ *C. d. eureka*
_____ 20. *Cicindela hirticollis*, Hairy-necked Tiger Beetle
_____ *C. h. hirticollis*
_____ *C. h. abrupta*
_____ *C. h. athabascensis*
_____ *C. h. coloradula*
_____ *C. h. corpuscula*
_____ *C. h. couleensis*
_____ *C. h. gravida*
_____ *C. h. rhodensis*
_____ *C. h. shelfordi*
_____ *C. h. siuslawensis*
_____ 21. *Cicindela limbata*, Sandy Tiger Beetle
_____ *C. l. limbata*
_____ *C. l. hyperborea*
_____ *C. l. labradorensis*
_____ *C. l. nogahabarensis*
_____ *C. l. nympha*
_____ 22. *Cicindela albissima*, Coral Pink Sand Dune Tiger Beetle
_____ 23. *Cicindela theatina*, Colorado Dune Tiger Beetle
_____ 24. *Cicindela arenicola*, St. Anthony Dune Tiger Beetle
_____ 25. *Cicindela waynei*, Bruneau Dune Tiger Beetle
_____ 26. *Cicindela columbica*, Columbia River Tiger Beetle
_____ 27. *Cicindela bellissima*, Pacific Coast Tiger Beetle
_____ *C. b. bellissima*
_____ *C. b. frechini*
_____ 28. *Cicindela formosa*, Big Sand Tiger Beetle
_____ *C. f. formosa*
_____ *C. f. generosa*
_____ *C. f. gibsoni*
_____ *C. f. pigmentosignata*
_____ *C. f. rutilovirescens*
_____ 29. *Cicindela purpurea*, Cow Path Tiger Beetle
_____ *C. p. purpurea*
_____ *C. p. audubonii*

_____ *C. p. cimarrona*

_____ *C. p. hatchi*

_____ *C. p. lauta*

_____ 30. *Cicindela ohlone*, Ohlone Tiger Beetle

_____ 31. *Cicindela pugetana*, Sagebrush Tiger Beetle

_____ 32. *Cicindela plutonica*, Alpine Tiger Beetle

_____ 33. *Cicindela splendida*, Splendid Tiger Beetle

_____ 34. *Cicindela denverensis*, Green Claybank Tiger Beetle

_____ 35. *Cicindela limbalis*, Common Claybank Tiger Beetle

_____ 36. *Cicindela decemnotata*, Badlands Tiger Beetle

_____ *C. d. decemnotata*

_____ *C. d. meriwetheri*

_____ *C. d. bonnevillensis*

_____ *C. d. montevolans*

_____ 37. *Cicindela sexguttata*, Six-spotted Tiger Beetle

_____ 38. *Cicindela denikei,* Laurentian Tiger Beetle

_____ 39. *Cicindela patruela*, Northern Barrens Tiger Beetle

_____ *C. p. patruela*

_____ *C. p. consentanea*

_____ 40. *Cicindela pulchra*, Beautiful Tiger Beetle

_____ *C. p. pulchra*

_____ *C. p. dorothea*

_____ 41. *Cicindela pimeriana*, Cochise Tiger Beetle

_____ 42. *Cicindela fulgida*, Crimson Saltflat Tiger Beetle

_____ *C. f. fulgida*

_____ *C. f. pseudowillistoni*

_____ *C. f. rumppi*

_____ *C. f. westbournei*

_____ 43. *Cicindela parowana*, Dark Saltflat Tiger Beetle

_____ *C. p. parowana*

_____ *C. p. platti*

_____ *C. p. wallisi*

_____ 44. *Cicindela scutellaris*, Festive Tiger Beetle

_____ *C. s. scutellaris*

_____ *C. s. flavoviridis*

_____ *C. s. lecontei*

_____ *C. s. rugata*

_____ *C. s. rugifrons*

_____ *C. s. unicolor*

_____ *C. s. yampae*

_____ 45. *Cicindela nigrior*, Autumn Tiger Beetle

_____ 46. *Cicindela tranquebarica*, Oblique-lined Tiger Beetle
     _____ *C. t. tranquebarica*
     _____ *C. t. arida*
     _____ *C. t. borealis*
     _____ *C. t. cibecuei*
     _____ *C. t. diffracta*
     _____ *C. t. inyo*
     _____ *C. t. joaquinensis*
     _____ *C. t. kirbyi*
     _____ *C. t. moapana*
     _____ *C. t. parallelonata*
     _____ *C. t. sierra*
     _____ *C. t. vibex*
     _____ *C. t. viridissima*
_____ 47. *Cicindela ancocisconensis*, Appalachian Tiger Beetle
_____ 48. *Cicindela lengi*, Blowout Tiger Beetle
     _____ *C. l. lengi*
     _____ *C. l. versuta*
     _____ *C. l. jordai*
_____ 49. *Cicindela tenuicincta*, Short-legged Tiger Beetle
_____ 50. *Cicindela latesignata*, Western Beach Tiger Beetle
_____ 51. *Cicindela willistoni*, Williston's Tiger Beetle
     _____ *C. w. willistoni*
     _____ *C. w. echo*
     _____ *C. w. estancia*
     _____ *C. w. funaroi*
     _____ *C. w. hirtifrons*
     _____ *C. w. praedicta*
     _____ *C. w. pseudosenilis*
     _____ *C. w. sulfontis*
_____ 52. *Cicindela senilis*, Senile Tiger Beetle
     _____ *C. s. senilis*
     _____ *C. s. frosti*
_____ 53. *Cicindela amargosae*, Great Basin Tiger Beetle
     _____ *C. a. amargosae*
     _____ *C. a. nyensis*

Genus *Cicindelidia*, American Tiger Beetles

_____ 54. *Cicindelidia nigrocoerulea*, Black Sky Tiger Beetle
     _____ *C. n. nigrocoerulea*
     _____ *C. n. bowditchi*

_____ *C. n. subtropica*

____ 55. *Cicindelidia hornii,* Horn's Tiger Beetle

____ 56. *Cicindelidia obsoleta,* Large Grassland Tiger Beetle

_____ *C. o. obsoleta*

_____ *C. o. neojuvenilis*

_____ *C. o. santaclarae*

_____ *C. o. vulturina*

____ 57. *Cicindelidia punctulata,* Punctured Tiger Beetle

_____ *C. p. punctulata*

_____ *C. p. chihuahuae*

____ 58. *Cicindelidia tenuisignata,* Thin-lined Tiger Beetle

____ 59. *Cicindelidia fera,* Red-lined Tiger Beetle

____ 60. *Cicindelidia highlandensis,* Highlands Tiger Beetle

____ 61. *Cicindelidia abdominalis,* Eastern Pinebarrens Tiger Beetle

____ 62. *Cicindelidia scabrosa,* Scabrous Tiger Beetle

____ 63. *Cicindelidia floridana* Miami Tiger Beetle

____ 64. *Cicindelidia politula,* Limestone Tiger Beetle

_____ *C. p. politula*

_____ *C. p. barbaraannae*

_____ *C. p. petrophila*

_____ *C. p. viridimonticola*

____ 65. *Cicindelidia cazieri,* Cazier's Tiger Beetle

____ 66. *Cicindelidia rufiventris,* Eastern Red-bellied Tiger Beetle

_____ *C. r. rufiventris*

_____ *C. r. cumatilis*

_____ *C. r. hentzii*

____ 67. *Cicindelidia sedecimpunctata,* Western Red-bellied Tiger Beetle

____ 68. *Cicindelidia melissa,* Melissa's Tiger Beetle

____ 69. *Cicindelidia ocellata,* Ocellated Tiger Beetle

_____ *C. o. ocellata*

_____ *C. o. rectilatera*

____ 70. *Cicindelidia marginipennis,* Cobblestone Tiger Beetle

____ 71. *Cicindelidia hemorrhagica,* Wetsalts Tiger Beetle

_____ *C. h. hemorrhagica*

_____ *C. h. woodgatei*

_____ *C. h. arizonae*

____ 72. *Cicindelidia schauppii,* Schaupp's Tiger Beetle

____ 73. *Cicindelidia sommeri,* Orange-banded Tiger Beetle (HYPOTHETICAL)

____ 74. *Cicindelidia trifasciata,* S-banded Tiger Beetle

_____ *C. t. ascendens*

_____ *C. t. sigmoidea*

Genus *Eunota*, Saline Tiger Beetles

___ 75. *Eunota californica*, California Tiger Beetle
  ___ *E. c. pseudoerronea*
  ___ *E. c. mojavi*
___ 76. *Eunota gabbii*, Western Tidal Flat Tiger Beetle
___ 77. *Eunota circumpicta*, Cream-edged Tiger Beetle
  ___ *E. c. circumpicta*
  ___ *E. c. johnsonii*
  ___ *E. c. pembina*
___ 78. *Eunota praetextata*, Riparian Tiger Beetle
  ___ *E. p. praetextata*
  ___ *E. p. pallidofemora*
___ 79. *Eunota fulgoris*, Glittering Tiger Beetle
  ___ *E. f. fulgoris*
  ___ *E. f. albilata*
  ___ *E. f. erronea*
___ 80. *Eunota pamphila*, Gulfshore Tiger Beetle
___ 81. *Eunota severa*, Saltmarsh Tiger Beetle
___ 82. *Eunota striga*, Elusive Tiger Beetle
___ 83. *Eunota togata*, White-cloaked Tiger Beetle
  ___ *E. t. togata*
  ___ *E. t. globicollis*
  ___ *E. t. fascinans*

Genus *Microthylax*, Coral Beach Tiger Beetles

___ 84. *Microthylax olivacea*, Olive Tiger Beetle

Genus *Habroscelimorpha*, Habro Tiger Beetles

___ 85. *Habroscelimorpha dorsalis*, Eastern Beach Tiger Beetle
  ___ *H. d. dorsalis*
  ___ *H. d. media*
  ___ *H. d. saulcyi*
  ___ *H. d. venusta*

Genus *Opilidia*, Opilid Tiger Beetles

___ 86. *Opilidia chlorocephala*, Lime-headed Tiger Beetle
  ___ *O. c. smythi*

Genus *Brasiella,* Little Tiger Beetles

____ 87. *Brasiella viridicollis,* Cuban Green-necked Tiger Beetle
____ 88. *Brasiella wickhami,* Sonoran Tiger Beetle
____ 89. *Brasiella viridisticta,* Pygmy Tiger Beetle

Genus *Parvindela,* American Diminutive Tiger Beetles

____ 90. *Parvindela terricola,* Variable Tiger Beetle
     ____ *P. t. terricola*
     ____ *P. t. cinctipennis*
     ____ *P. t. continua*
     ____ *P. t. imperfecta*
     ____ *P. t. kaibabensis*
     ____ *P. t. susanagreae*
____ 91. *Parvindela lunalonga,* Meadow Tiger Beetle
____ 92. *Parvindela lemniscata,* White-striped Tiger Beetle
     ____ *P. l. lemniscata*
     ____ *P. l. rebaptisata*
____ 93. *Parvindela debilis,* Grass-runner Tiger Beetle
____ 94. *Parvindela celeripes,* Swift Tiger Beetle
____ 95. *Parvindela cursitans,* Ant-like Tiger Beetle

Genus *Apterodela,* Leaf Litter Tiger Beetles

____ 96. *Apterodela unipunctata,* One-spotted Tiger Beetle

Genus *Dromochorus,* Dromo Tiger Beetles

____ 97. *Dromochorus pruininus,* Frosted Tiger Beetle
____ 98. *Dromochorus belfragei,* Loamy-ground Tiger Beetle
____ 99. *Dromochorus knisleyi,* Juniper Grove Tiger Beetle
____ 100. *Dromochorus welderensis,* Gulf Prairie Tiger Beetle
____ 101. *Dromochorus chaparrelensis,* Chaparral Tiger Beetle
____ 102. *Dromochorus minimus,* Pygmy Dromo Tiger Beetle
____ 103. *Dromochorus velutinigrens,* Velvet Tiger Beetle
____ 104. *Dromochorus pilatei,* Cajun Tiger Beetle

Genus *Ellipsoptera,* Ellipsed-winged Tiger Beetles

____ 105. *Ellipsoptera hamata,* Coastal Tiger Beetle
     ____ *E. h. monti*
     ____ *E. h. lacerata*

____ 106. *Ellipsoptera marginata*, Margined Tiger Beetle

____ 107. *Ellipsoptera blanda*, Sandbar Tiger Beetle

____ 108. *Ellipsoptera wapleri*, White-sand Tiger Beetle

____ 109. *Ellipsoptera nevadica*, Nevada Tiger Beetle

    ____ *E. n. nevadica*

    ____ *E. n. citata*

    ____ *E. n. knausii*

    ____ *E. n. lincolniana*

    ____ *E. n. makosika*

    ____ *E. n. olmosa*

    ____ *E. n. tubensis*

    ____ *E.* n. ssp.

____ 110. *Ellipsoptera cuprascens*, Coppery Tiger Beetle

____ 111. *Ellipsoptera macra*, Sandy Stream Tiger Beetle

    ____ *E. m. macra*

    ____ *E. m. ampliata*

    ____ *E. m. fluviatilis*

____ 112. *Ellipsoptera puritana*, Puritan Tiger Beetle

____ 113. *Ellipsoptera sperata*, Rio Grande Tiger Beetle

    ____ *E. s. sperata*

    ____ *E. s. inquisitor*

____ 114. *Ellipsoptera marutha*, Aridland Tiger Beetle

____ 115. *Ellipsoptera lepida*, Ghost Tiger Beetle

____ 116. *Ellipsoptera gratiosa*, Whitish Tiger Beetle

____ 117. *Ellipsoptera hirtilabris*, Moustached Tiger Beetle

Index

CPSIA information can be obtained
at www.ICGtesting.com
Printed in the USA
LVHW080724020222
710044LV00012B/248

9 780199 367177